EAGLES OF THE LUFT

FOCKE-WULF Fw 200
MATTHEW WILLIS

TEMPEST
BOOKS

First published in Great Britain in 2023
by Tempest Books
an imprint of Mortons Books Ltd.
Media Centre
Morton Way
Horncastle LN9 6JR

www.mortonsbooks.co.uk

ISBN 978 1 911658 65 8

The right of Matthew Willis to be identified as the author of this work has been asserted in accordance with the
Copyright, Designs and Patents Act 1988.

All images author's collection unless otherwise stated

Typeset by Jayne Clements (jayne@hinoki.co.uk), Hinoki Design and Typesetting

Printed and bound in Great Britain

Acknowledgements
The author is indebted to Anke Tymens, Giulio Poggiaroni, Lynn Ritger, Paul Beaver and Dan Sharp.

CONTENTS

1

1936—39

ORIGIN

On 19 January 1936, the Reich Air Ministry (RLM) issued a request for a four-engined airliner to meet a requirement of the Deutsche Luft Hansa (DLH) airline. DLH's standard passenger aircraft at the time was the venerable Junkers Ju 52/3m, of which it operated more than 200 at the height of the trimotor's career. By the mid-1930s however, a new generation of streamlined, high-performance monoplanes was appearing and making aircraft like the Ju 52 seem slow, dated and uncomfortable. DLH even diverged from its usual practice of buying only German aircraft when it acquired Boeing 247 and Douglas DC-2 airliners in 1935.

DLH had developed transatlantic routes in the 1930s with flying boats, developing Germany's links with its expatriate communities and business interests in South America. These aircraft were rather cumbersome, however, involving refuelling by ship at sea and launch by catapult to maximise endurance. Flights could take several days.

According to legend, Kurt Tank, chief designer of Focke-Wulf Flugzeugbau (FWF), bumped into Dr.-Ing Rudolf Stüssel, Technical Director of DLH while waiting for a train in January 1936. Having seen that aircraft like the DC-2 were the future, Stüssel was on the lookout for a modern machine from the domestic aircraft industry. Tank, it is said,

insisted that FWF could do better, that such a machine, a modern landplane, could, by means of the efficiency that was now possible, make direct transatlantic flights. This would vastly simplify the business of transatlantic passenger, freight and mail flights as well as cutting the time.

Just days after the meeting supposedly took place, the RLM issued DLH's request to the industry. It is not clear whether Tank's boast about the possibility of a landplane that could regularly undertake transatlantic flights is true, or whether it influenced either DLH or the RLM. This would have represented a huge advance on current practice and questionably achievable with current technology. In any case, it seems that Stüssel was more interested in an aircraft that could replace the Ju 52 on European routes.

Regardless of the ultimate motivation behind the new airliner, there was surely little doubt that Focke-Wulf was in prime position to fulfil the airline's requirement, despite the company's lack of experience with large commercial aircraft. Much of the industry was busy with the rearmament of the Luftwaffe, which had been announced in March 1935 in defiance of the Versailles Treaty, and this had also been the priority for the RLM, leaving little attention given to commercial aircraft. It was painfully apparent within a short time that

The first Fw 200, V1 D-AERE *Brandenburg*, on rollout, showing the original close-fitting cowlings, unswept outer wing and single-wheel undercarriage.

Germany would fall far behind in this sector unless rapid attention was paid to the development of new airliners. Other than the small Heinkel He 70 Blitz mainplane, there were no fully modern German commercial aircraft in service in 1935, and few genuine airliners in prospect. While Heinkel, Junkers and Dornier were ostensibly working on modern commercial aircraft, these were in reality primarily military projects, and their small capacity and low comfort for passengers meant they could not be regarded as viable competitors to the newest American designs or realistic replacements for the Ju 52. DLH had approached Junkers and Dornier, which were both working on four-engined long-range bombers, for a dedicated airliner but without much success.

FWF's bread and butter during the 1920s had been airliners, hence Stüssel's familiarity with Tank, but these were mostly small and of traditional design. After merging with Albatros in 1931, (gaining Tank into the bargain), FWF adapted to all-metal structures, and by 1935 was producing partially stressed-skin designs that were conceptually up to date. The Fw 58 Weihe, which first flew in January 1935, was a modern low-wing, twin-engined monoplane with retractable undercarriage developed as a multi-engine aircrew trainer for the Luftwaffe. It was not a fully cantilever monoplane, with the wing and tailplane each braced by a single strut on either side of the fuselage, but otherwise the Weihe was of clean and efficient appearance. The Fw 57 heavy fighter flew the same year; it had true cantilever flying surfaces and was of clean, duralumin-skinned construction but lost out to the Messerschmitt Bf 110, leaving FWF with capacity on its hands at the Bremen factory and design office. The specification was finalised on the 26 February.

Tank delegated the task of realising his scheme to Oberingenieur Ludwig Mittelhuber. He was apparently clear that the airliner should have a slim fuselage to keep frontal area down and long, high aspect-ratio wings for efficiency. It was said that Tank was influenced in this by his experience of designing gliders in the early 1920s. High aspect-ratio (the ratio of span to chord) was identified by Tank as a way of increasing speed on relatively low power. This was confirmed by aerodynamic investigations: "In the case of low-powered airplanes, however, as, for example, the recent high-speed transport planes of the type Fw 200 or the Douglas DC4 ... a considerable gain is attainable through increase in the aspect ratio ... The increase in speed thus gained is the more marked the greater the aerodynamic efficiency of the aeroplane."[1]

Those wings were the subject of considerable work by Focke-Wulf, as they were vital to the success of the aircraft. It was essential that they were as efficient in reality as on paper, but it was equally important that they offered good control and had no unpleasant or unsafe characteristics. A programme of testing in the Focke-Wulf wind tunnel aimed to establish these parameters and fine-tune the form, although this had to be carried out quickly because of the tight schedule imposed on the design process (allegedly tightened further by Tank making a bet that FWF could have a prototype flying within a year of an order being placed).[2]

Focke-Wulf's favoured wing planform in the mid-1930s was a straight taper to a narrow, rounded tip, and usually a short, straight centre section and dihedral on the outer wings. American NACA (National Advisory Committee for Aeronautics) aerofoils were preferred. However, the company had not come to a consensus on the sweep of the lift-line: the Fw 58 had a swept leading edge with a straight trailing edge leading to a slightly swept line of maximum lift; the Fw 57 had a taper on both leading and trailing edges; the Fw 187 fighter developed around the same time as the new airliner had a virtually straight leading edge and forward swept trailing edge. It is no surprise therefore that during wind-tunnel testing for the airliner, probably undertaken in early 1936, variations on the planform were tried out to establish which of these offered the best performance and the least degradation in control at high angles of attack and yaw.

G. Hartwig, an aerodynamicist with FWF, published the findings of this programme in 1941, describing it thus: "In the course of development of the Focke-Wulf Fw 200 (Condor) wind tunnel tests were made on the breakdown of flow on several wings of identical chord distribution but dissimilar course of the lifting line, and extended on one wing to the case of appended fuselage, engine nacelles and flaps deflected."[3]

Three different planforms were tested related to the wing designed for the new airliner, one with straight leading edge and swept forward trailing edge, one straight trailing edge and aft-swept leading edge, and a third with equal taper. The first two were simple flat wings with no dihedral or twist. The last, demonstrating that it was already the preferred layout for the new airliner, had dihedral on the outer sections and twist ranging from 2.2° positive incidence at the centre section to 0.6° negative incidence at the tip. This arrangement

---

1   B. Göthert, Effect of Wing Loading, Aspect Ratio, and Span Loading on Flight Performances, *Deutsche Luftfahrtforschung* Vol. 16, No. 5, 20 May 1939, Verlag von R. Oldenbourg, München und Berlin, p.14

2   J. Scutts, *Focke-Wulf Fw 200 Condor*, Crécy, 2010 p. 15.

3   G. Hartwig, Observations of the effect of wing appendages and flaps on the spread of separation of flow over the wing, *Deutsche Luftfahrtforschung* Vol. 18. Nos. 2–3, 29 March 1941, pp. 40–46.

D-AERE in flight after modifications to its cowlings and vertical tail but with the original wing.

was intended to ensure that the tip stalled last, retaining attached airflow over the ailerons as late as possible, and that the inner part of the wing stalled early, imparting, it was hoped, an automatic stall warning to the pilot by gentle buffeting when airflow began to break down. The equal-taper wing was built with articulating ailerons and flaps and could also be fitted with representations of the engine nacelles and a section of fuselage. The wings with straight leading or trailing edge were not tested with twist, dihedral or appendages due to time constraints.

"The models were made of well-polished gypsum over steel-zinc framework." They were 1.320 metres in length, and carefully attached with fine silk tufts to indicate airflow breakdown. Hartwig remarked that few tests of this nature had been carried out and "there is little data on the breakdown of flow over the wing in the presence of appendages such as fuselage, engine nacelles, or deflected flaps

within the practical range of spans".

The tests initially threw up worrying results. The characteristics of the straight-edged wings were largely similar but all the wings, even the one with dihedral and twist, exhibited unfavourable breakdown of airflow, which contrary to hopes, broke down first towards the ends of the wings and at the trailing edge — the worst-case scenario for control. This was described by Hartwig as a "quite hopeless result".[4] Interestingly, the situation changed dramatically when engine nacelles and fuselage were added. Suddenly, the wing began behaving as anticipated and exhibiting much better characteristics for lateral control.

This no doubt reassured Tank and his team, and while the design of the wing continued to evolve — the aerofoil changed from a NACA 2218 to a NACA 2318 profile — the pattern was largely set. (In fact, the planform would change due to necessity after the first prototype flew when, due

---

4   Ibid. p. 45.

A series of four photographs showing D-AETA Westfalen during its politically important route-proving and flag-showing flights during May 1938 for DLH. The aircraft is pictured here during a visit to Latvia.

to centre of gravity concerns, the outer wings were swept back slightly. It seems likely that the wind-tunnel tests outlined above would have given some confidence to FWF that this could be carried out without any significant degradation in handling or control.)

Further tests were carried out including a 1:16 'reflection plane' model of part of the wing and inboard engine nacelle. "The test conditions included cruising flight, climbing flight and descending flight with the propeller operating as a brake. Both the nacelle and the wing were extensively instrumented with surface pressure taps."[5] The tests assessed the wing's lift in a variety of conditions and how it was affected by the presence of the nacelle and propeller.

As mentioned, the design phase and the testing associated with it was carried out quickly. Nevertheless, the fact that wind-tunnel testing to influence the design, a relatively rare occurrence in the mid-1930s, was carried out at all was innovative and indicated the careful and modern approach with which Tank and his team approached the design of the new aircraft.

The new airliner was schemed to carry up to 30 passengers for short-haul flights and 17 in considerable comfort during long-haul flights. There were to be two separate cabins, smoking and non-smoking. The crew would comprise four personnel — two pilots, a navigator and a steward. A galley was provided so passengers could have hot meals.[6] The design cruising altitude was 3,000 metres (10,000 feet), as unlike the Boeing 247, the cabin was not to be pressurised.

Discussions between DLH and FWF proceeded until June, while design work progressed; a full-scale mock-up of the fuselage was presented, which both DLH and the RLM approved. On 9 July, FWF provided the RLM with Tender No. 760,

and on 13 August, the RLM placed a development contract with FWF, according to which the company would produce two prototypes,[7] designated V1 and V2 (for *Versuchsmuster*, or experimental model).

The RLM had quickly spotted the potential of the new airliner for boosting the prestige of the Third Reich. Thus, from the beginning, the aircraft would be more than just a cutting-edge commercial aircraft but a means of advancing the Reich in what would today be termed 'soft power'. For that reason, they allocated the designation Fw 200, skipping ahead from FWF's approved block of designations (which by this time had reached 187) to afford a suitably memorable number.

Furthermore, the name Condor, after the South American raptor, was approved. This followed FWF's standard policy of naming aircraft after birds, but the choice of a large South American raptor represented a statement — more than one, in fact. Apart from the obvious parallels between the long-winged, soaring bird and its namesake, the name reinforced Germany's links with the South American continent and the Reich's global aspirations (indeed, a German–Brazilian airline called Syndicato Condor had been operating in South America since 1927). It also hinted at the transatlantic potential of Reich air travel, though it would have been clear even at this stage that the Fw 200 would not be capable of operating regularly as a transatlantic passenger aircraft.

THE CONDOR TAKES WING

Design of the Fw 200 was delegated to Andreas von Faehlmann, who had joined FWF in May 1933 as head of the design department. Weight-saving was of particular importance, as demonstrated by lubricant tanks being constructed from Elektron magnesium alloy with no additional protection.

5   W. Albring, Pressure Distribution Measurements of a Wing Model with Engine Nacelle and Rotating Propeller (Druckverteilungsmessungen an einem Tragflugelmodell mit Motorgondel und laufendem Propeller), *Deutsche Luftfahrtforschung*, Forschungsbericht FB 1908, May 1942.
6   Focke-Wulf Condor – die fliegende Schönheit, *Abenteuer Luftfahrt-Geschichte FliegerRevue X.*
7   G. Ott, Focke-Wulf Fw 200 "Condor" Teil 1 – *Luftverkehr, Transport und Schulung*, AirDOC 2007, p. 3.

The second Fw 200, V2 D-AETA *Westfalen*, in its original form with taller, more tapered vertical tail.

Tests were carried out in July and August 1937 to ensure the tanks were sufficiently safe in the event of an engine fire or forced landing.[8]

Construction of the V1 and V2 was overseen by Dipl.-Ing Wilhelm Basemir and progressed swiftly through 1937. Well before the V1 was completed, on 5 April, DLH authorised FWF to acquire materials for three more Fw 200s, although the airline would not be required to purchase them if the prototypes did not meet the required performance.[9] The first prototype, Werk-nummer (W. Nr.) 2000, was completed by the end of August and made its first flight, piloted by FWF chief test pilot Hans Sander with Kurt Tank in the second pilot's seat, on 6 September. This went smoothly and manufacturer's trials proceeded apace. Due to delays with the supply of BMW 132 G engines, the V1 was fitted with Pratt & Whitney S1EG Hornet

9-cylinder air-cooled radial engines. (In fact, the BMW was a German development of the Hornet.) Testing highlighted some shortcomings with the engine installation, particularly relating to the snug-fitting NACA-style cowlings. A report from the DLH technical department noted: "Testing of the FW-200 V1 and V2 did not confirm the expected progress. Initially, it was difficult to maintain the normal temperature of the engine cylinders, which led to the need to change the shape of the engine cowlings."[10]

On 27 November, the V2, W. Nr. 2484 was rolled out, fitted with BMW 132s from the start. The following month, the two prototypes were presented to the public during an event at Berlin-Tempelhof airport. Following manufacturer's trials, the V1 was given the civilian registration D-AERE and the V2 D-AETA. It was DLH

8 Bundesarchiv file R/3/3706.
9 G. Ott, Condor – from Airliner to Missile Carrier, *Classic Wings 101* Vol. 22 No. 3, 2015, p. 48.
10 История самолета #3 (*History of the Aircraft #3*) – Fw 200 Condor, 2006, pp. 20–21.

practice to name aircraft after cities or regions of Germany, and consequently D-AERE was named *Brandenburg* and D-AETA *Westfalen*. By the end of 1937, the V1 had flown around 90 times, and the V2 around 60.[11] DLH had seen enough, and not only did the airline give the go-ahead to the three initial production aircraft, but brought the order up to six, which were dubbed the A-0 model by the RLM.

The early test flying had proved the success of the design, which was not to say that the aircraft was perfect. Alterations were made to the tail surfaces, when the fin and rudder as designed proved to have insufficient area, and a tendency of the elevators to overbalance was addressed too. A taller, more tapered fin and rudder was introduced from the beginning on D-AETA, but the solution adopted for all Fw 200s was a fin and rudder of the same height as the original design but of overall broader chord and less taper, making for a distinctively square appearance, together with longer span and less tapered tailplane and elevators. Even this did not realise the benefits hoped for. The DLH technical department complained of "The lack of stability relative to the transverse axis when heavily loaded," adding "The changes carried out in the vertical tail have not yet brought the expected effect, and as a result, further work is necessary."[12]

In the face of marginal stability, and perhaps a recognition that there was too little flexibility in the centre of gravity (C/G) for future weight increases, FWF instigated modifications to the wing. The ideal solution would have been to sweep the entire wing back, but this would have involved significant changes to the design of the inner wing, including the engine nacelles with fuel system, not to mention any jigs and tooling already under construction. The simplest fix was to sweep the outer wing panel where it attached to the centre wing section just outboard of the outer

engine nacelles. This led to a distinctive kink in the trailing edge and a 'cranked' appearance to the wing when viewed in plan but solved the stability problem. It had the effect of marginally reducing the span (from 32.97m to 32.84m) and area (from 120.00m2 to 118.00m2).

The first Fw 200 that would appear with the definitive flying surfaces was therefore the first of the initial series A-0 aircraft, designated the S1 by FWF and registered D-ADHR with the name *Saarland* (W. Nr. 2893). The prototypes had continued the testing programme, with the V1 undertaking DLH's 200-hour evaluation before being returned to FWF in February for modification to bring it closer to A-0 standard. D-AETA meanwhile set out on a series of route-proving flights for DLH in May, visiting sites across Europe with much attendant publicity.

The Fw 200 had not yet gone into airline service and yet FWF had received its first external orders — in March, the Danish airline Det Danske Luftfartselskab (DDL) purchased two Fw 200 A-0s. In fact, the two aircraft for DDL were taken from the initial series machines, the S2 (W. Nr 2894) and S4 (W. Nr 2993), and orders for two further A-0s were added to complete DLH's requirement. The two Danish aircraft were redesignated KA-1, this essentially being the export variant of the A-0.

Discussions with other airlines were ongoing too. British Airways Ltd (BAL) had an existing relationship with DLH, running a night mail service from London to Cologne and Hanover in partnership with the German airline. BAL was looking for an aircraft for South American routes, and as this was on the agenda for DLH too — Syndicato Condor, DLH's partner airline in Brazil, would later operate two Fw 200 A-0s — it made sense to consider the same type. Negotiations between BAL and FWF sales manager Hans Junge probably began in late 1937[13] having been permitted by

11 G. Ott, Condor — from Airliner to Missile Carrier, *Classic Wings 101* Vol. 22 No. 3, 2015, p. 49.
12 История самолета #3 (*History of the Aircraft* #3) — Fw 200 Condor, 2006, pp. 20–21.
13 B. Wheeler, Focke-Wulfs in Britain, Q&A, *Aeroplane*, March 2018, p. 20.

A sign of things to come. A Luftwaffe *Oberfeldwebel* poses in front of D-ADHR Saarland while still in DLH service between July 1938 and September 1939. Saarland would be one of the first Fw 200s to go into military service.

the Air Ministry. BAL's managing director Major J. R. McCrindle visited FWF's Bremen factory on 21 January 1938, and a week later a full technical team from the airline followed under the leadership of test pilot Alan Campbell-Orde. The team were allowed to scrutinise D-AERE and offered a version of the Fw 200 A tailored to BAL's requirements. In June, D-AETA flew to Croydon for inspection by BAL. In the end, the airline did not pursue the Fw 200 – Campbell-Orde felt the cabin to be narrow, suggesting that it reflected military priorities, and expressed concern that the high-aspect ratio wing would actually limit range rather than the opposite.

FWF was not about to let this setback affect their ambitions. The first series aircraft, D-ADHR, was completed in June and the opportunity seized for a major publicity coup. Tank proposed to carry out an ambitious Berlin–Cairo–Berlin flight within 24 hours, with passengers on board, in

the hope that a new record could be set. The Fw 200 took off on 27 June and the outward leg was completed without incident, but the attempted achievement failed when the tailwheel broke during a landing at Salonica and could not be repaired in time. D-ADHR was delivered to DLH on 23 July 1938.

After the Berlin–Cairo–Berlin disappointment, FWF planned an even more ambitious flight to follow up, such was the company's confidence in the aircraft. The V1 returned to flight after its modifications in July, with the new and particularly appropriate registration D-ACON. The publicity flight would entail nothing less than a circumnavigation of the globe, again with hopes of a record, and begin with a non-stop flight from Berlin to New York! Thus were the transatlantic credentials of the Condor to be given a polish, but in reality, the flight could not be made with the aircraft in passenger trim. The passenger cabin

Fw 200 S3 D-AMHC Nordmark was the second production Fw 200, entering DLH service in August 1938. EDWIN HOOGSCHAGEN

was stripped of anything that could be removed and five additional 1,100-litre fuel tanks installed in the space. Trial flights were undertaken in early July and the transatlantic flight scheduled for the following month.

Here, geopolitics intervened as the US authorities declined to give permission for the necessary overflights and landings. This meant the abandonment of the circumnavigation attempt but with a degree of subtlety, the transatlantic flight could go ahead. DLH had already been given permission to operate up to 28 non-commercial evaluation flights to the US across the Atlantic during 1938. The planned flights were to have been carried out with Blohm & Voss Ha 139 flying boats, but the type of aircraft had not been specified, so

DLH nominated the Fw 200 flights to and from the US as two of the permitted evaluations. The outbound flight, with DLH Kapitän Alfred Henke at the controls, was completed without a hitch and D-ACON arrived at Floyd Bennett Field at 1.50pm on 11 August, having taken 24 hours and 56 minutes since taking off from Berlin-Staaken and establishing a new record. Minor damage caused by bad weather was repaired over the next few days and D-ACON returned to Germany on 13 August, shaving five hours and one minute off the outbound time. Interestingly, the crew included co-pilot Rudolf Freiherr von Moreau, a Luftwaffe *Hauptmann*.

The deteriorating political situation again interrupted Tank's efforts to publicise his airliner with impressive endurance flights. A plan to fly D-ACON from New York to Tokyo had to be postponed due to negotiations over Sudetenland raising international tensions. Indeed, this period saw the first military adoption of the Fw 200 when D-ACON — which was owned by the RLM — was temporarily assigned to the Luftwaffe Lehrgeschwader (training wing) at Jüteborg as a transport aircraft during September, at the height of what became known as the Munich Crisis.

The settlement of that situation caused tensions to subside for the time being, and towards the end of November, with D-ACON now released from Luftwaffe duty, the flight east was rescheduled. The same crew that had flown D-ACON to New York took the V1 in the opposite direction on 28 November 1938, reaching Tachikakawa Airport, Tokyo, in 48 hours and 18 minutes, representing another long-distance speed record. This flight was accompanied by a sales push by FWF to sell the Fw 200 to Japanese airlines. The flight gained the attention of the Manshū Kōkū Kabushiki-gaisha (MKKK, the Manchuria Air Transport Company), and FWF's head of sales, Heinz Junge, quickly entered negotiations.

An embarrassing incident on the return flight threatened to stymie FWF's sales offensive, however. D-ACON ditched just off the coast of Manila after apparently suffering fuel starvation and the failure of the two starboard engines while on final approach to Nielson Field. Tank attributed this to an error on the part of the flight engineer in operating the fuel cross-feed incorrectly and angrily blamed the captain and crew for the damage to the aircraft and company's reputation. In fact, the RLM's subsequent investigation exonerated the crew, but the damage was done, and Dutch airline KLM withdrew from an option to purchase nine Fw 200s. To make matters worse, D-ACON suffered irreparable damage while being recovered. An additional aircraft for DLH was

ordered to replace it.

Despite the incident, on 17 December MKKK confirmed that it would be purchasing five Fw 200s, so all was far from lost.

INTO AIRLINE SERVICE

DLH received two initial series Fw 200s in 1938, S1 D-ADHR *Saarland* in July as mentioned, and S3 D-AMHC *Nordmark* in August. Along with D-AETA, these aircraft flew more than a quarter of a million kilometres in the first year of service and proved popular and reliable. One innovation DLH introduced on the Fw 200 was a steward on each flight.[14] This was successful enough that DDL replicated it on its own Fw 200 service. Once the route-proving flights had been completed, D-AETA was returned to FWF for conversion to the latest standard, and thereafter retained by the RLM.

The two Danish Condors were introduced to DDL service in the second half of 1938. The S2, which was registered OY-DAM and named *Dania*, joined the airline on 14 July, followed on 15 November by the S4, OY-DEM *Jutlandia*.

DLH received two further Fw 200 A-os from the initial series in 1939 for its own service, D-ARHW *Friesland* and D-ABOD *Kurmark*, in January and August, and took on two more only to transfer them to Syndicato Condor in Brazil, D-AXFO *Pommern* becoming PP-CBI *Abaitará* in July with D-ASBK *Holstein* being re-registered PP-CBJ and given the name *Arumani* in August. The South American aircraft were flown to Brazil by crews made up of DLH and Syndicato Condor personnel.

Further sales were being struck by FWF, helped by DLH's partnership agreements with overseas airlines. Another four machines were ordered by the Chinese government in July for the Eurasia airline, which DLH had an interest in, though this led to a delicate political position with Japan, which had already bought Condors and was at war with China. RLM State Secretary Erhard Milch

---

14 G. Ott, Focke-Wulf Fw 200 "Condor" Teil 1 — *Luftverkehr, Transport und Schulung*, AirDOC 2007, p. 6.

wrote to Göring in July outlining the situation:

> The Chinese government intends to buy 4
> Focke-Wulf (Condor) airliners for Eurasia,
> two of which are to be delivered in Febru-
> ary 1940 and two in August 1940. It is clear
> that these aircraft are only to be used for
> service with Eurasia. Lufthansa, which
> has so far managed to ensure that only
> German aircraft are used in the operation
> of Eurasia, of which it owns 1/3 of the
> shares, is urgently asking for permission,
> because otherwise American aircraft will
> be bought ... According to the instructions
> available, the deal could be approved, as
> it is not military equipment. However, the
> report was made to Field Marshal because
> Japan had long opposed the operation of
> the Eurasia, which even today is still being
> flown by German pilots. Presumably Japan
> will claim that the Eurasia's operations are
> of indirect military importance because
> they provide the only connection (includ-
> ing communications) between China and
> the coast without Japanese control. Never-
> theless, with reference to the purely civilian
> character of the device, the approval of the
> business is proposed, since the Ministry
> of Economics attaches great importance
> to its implementation and the consent of
> Japan is unlikely to be obtained.[15]

The plan was to ship the Condors to Hong Kong
and assemble them there.

A further Fw 200 was allocated for government
service, the S8 D-ACVH *Grenzmark*, in the summer
of 1939 while the S9 was earmarked as Hitler's
personal transport (see below).

As the first series of Condors went into service,
FWF continued to evolve the design, and by the
end of 1938 was planning a significantly upgraded
model, the Fw 200 B. This represented a relatively
substantial development of the type from the
A-0. The all-up weight was to be increased from
32,000lb to 38,000lb, although only minor airframe
strengthening was carried out.[16]

The most obvious external alteration was a
complete redesign of the engine installation. The
engines themselves were the 870hp H model of
the BMW 132, replacing the 720hp G model used
on the A-0 aircraft.

The nacelles and cowlings were of entirely
new form, with the inner nacelle being extended
forward, moving the engine further ahead of the
outer engine than on the preceding models. This
was likely done chiefly for reasons of balance,
especially considering the increased payload
weight. It may also, however, have helped to
reduce interaction between airflows from each
propeller. The cowlings were slightly squarer in
profile and longer, while moveable cooling gills
were now present on the aft end. This model was
to be the first to use three-bladed propellers.

The undercarriage was revised, replacing the
single main wheel on each unit with twin, smaller
wheels, setting the pattern for all future Fw 200s.

As many as 46 B model Condors were ordered
by airlines by August 1939, in slightly different
versions suiting each potential operator. The Finnish airline Aero O/Y which purchased two (receiving the internal Focke-Wulf designation KB-1 and
the W. Nrs. 0009 and 0010), while as mentioned
above, Japanese airline MKKK placed orders for
five (designated KC-1 by Focke-Wulf, with the W.
Nrs. 0017 to 0021). The Japanese order was initially
signed in December and reported in the media
as early as 8 January. MKKK intended to use the
Condors for long-range continental routes across
the puppet Manchoukuo state. According to a
British intelligence report:

> The machine body will be the same as

15   E. Milch, Memo to Reichsmarschall Göring, Subject: Delivery of 4 Fw 200 (Condor) aircraft for the Eurasia, July 1939.
16   FW 200 Aircraft, Air Ministry, Directorate of Intelligence and related bodies, Intelligence reports and papers, AIR 40/154.

The V1 as rebuilt with the definitive swept outer wing, reregistered D-ACON, as it undertook the record-breaking flight to New York in August 1938.

the one which paid a visit to Japan, the BMW 132 L type equipped with four 870 hp motors and designed to accommodate 26 passengers. The Japan Aviation Corporation, however, is going to have the planes altered to accommodate only nine persons; while the remaining accommodation, for 17 persons, will be taken up by the addition of a tank, which will enable the planes to fly continuously for a distance of 4,500 kilometres. Thus, with Tokyo as a starting point, they may fly west through Peking, Suiyuen and to Sian. In a south westerly course they may fly past Hanoi, Bangkok to Rangoon or Akyab. Flying south they may reach Borneo or New Guinea. In the Pacific Ocean they may reach Midoyei. Finally in a northern course they may, after calling at Hokkaido, fly straight on non-stop to Nome … The Japan Aviation Corporation is picking more than 20 accomplished pilots, mechanics and wireless operators to send them to Germany for further training.[17]

Reports were even received that some of the Condors had arrived at Hainan in September 1940, but it's most likely this was a case of mistaken identity with Junkers Ju 86s operated by MKKK.

Efforts were underway as late as May 1940 to deliver the Chinese aircraft to Eurasia. A British Air Ministry intelligence report noted: "One of our sources has been informed that Walter LUTZ, pilot

17   Purchase of 'German Planes', 8 January 1939, AI Report No. 23629, dated 27.4.39, Air Ministry, Directorate of Intelligence and related bodies, Intelligence reports and papers, AIR 40/154.

D-ACON's passenger cabin, showing the steward – Fw 200s were the first aircraft in scheduled operation to introduce this service.

in the EURASIAN air service will leave shortly for Germany in connection with the purchase of four CONDOR planes for the CHUNKING – ALMA ATA route." $220,000 was reportedly raised to cover the purchase of the aircraft through an intermediary of the China Air Motive Company in Canada.[18]

Despite the large number of orders, however, and pressure within Germany to continue with overseas sales even after the outbreak of war, no Fw 200 B airliner would see the light of day.

THE CONDOR GOES TO WAR

The first Fw 200 in military service was the V1 during its brief allocation to the training squadron at Jüteborg during the Munich Crisis, which presaged the destiny of the existing Fw 200 fleet.

By the summer of 1939, steps were already being taken to make the most of the Fw 200's performance, particularly its endurance, for military applications in a dedicated version rather than simply as a temporary military transport.

Oberstleutnant Theodor Rowehl had pioneered strategic reconnaissance in the 1930s, and in 1936 was placed in charge of a covert photographic reconnaissance unit within the Luftwaffe, the cryptically named Fliegerstaffel zur besonderen Verwendung (z.b.V) or Squadron for Special Purposes, later changed to Versuchsstelle für Höhenflüge or VfH (High-Altitude Flight Research Department) but always known colloquially as Fliegerstaffel Rowehl. By the middle of 1939 it was a question of when rather than if war with the

---

18   Extract from AI774 – Information dated 8.5.1940, AI2c, Ref. P/287/40, Air Ministry, Directorate of Intelligence and related bodies, Intelligence reports and papers, AIR 40/154.

"Vergebens, mein Lieber!
Wir fliegen auch mit zwei Motoren ruhig weiter."

A big selling point of the Fw 200 for passengers was its ability to fly safely with some of its engines shut down. This promotional postcard declares, "In vain, my friend! We can fly just as well on two engines!" despite the devil's best efforts.

United Kingdom would break out, and the unit had no aircraft with sufficient range to reconnoitre the Royal Navy's fleet anchorage at Scapa Flow. The V2 D-AETA, owned by the government, was taken up for modification to long-range photographic reconnaissance under Rowehl's technical director Fritz Klemm. A DLH pilot, Flugkapitän Franz Klaus, was attached to the unit to instruct crews. In late August, the S3 D-AHMC *Nordmark* was appropriated from DLH to undergo a similar modification.

In the meantime, at Bremen, production of the B model airliner, of which 46 had been ordered, was underway. In late August the V4 prototype Fw 200 B (W. Nr. 0001) was briefly completed in its civil guise and was shortly to be transferred to DLH when it, too, was earmarked for Rowehl's reconnaissance unit. The aircraft was redesignated V10 to indicate that it would be the prototype of a distinctly different version, and it re-entered the factory for rebuild. The modifications to 0001 would be more extensive than those applied to the V2 and S3, and the V10 became the first armed Fw 200, with defensive turrets and a ventral *Rumpfwanne*, or fuselage tub.[19] These innovations, albeit in a somewhat different form, would mark

19 This structure is typically referred to as the gondola in English sources, due to its similarity to an airship's gondola. However, as this conflicts with the German term for engine nacelles ('*gondeln*'), this text will use the term 'tub' as a closer translation and to avoid confusion.

out the majority of military Condors thereafter. It would be delivered on 14 November 1939.

The remaining B models under construction were appropriated for the Luftwaffe, and all overseas orders cancelled on the outbreak of war. To mark the difference from the Fw 200 B, new designations were applied relating to the roles the airframes were completed for – Fw 200 C for armed, combat aircraft and Fw 200 D for unarmed military transports. Three of the five Japanese Bs emerged as Fw 200 Cs, while the remaining two and the two Finnish aircraft were completed to Fw 200 D specification.

The general shortage of transport aircraft in the Luftwaffe, particularly modern ones of relatively good performance, meant that DLH's Condors were in demand as Germany was gearing up for war in late August 1939.

Several of the airline's four-engined aircraft were appropriated as military transports and assembled in a special unit, Kampfgeschwader zur besonderen Verwendung 172 (Combat Group for Special Purposes) or KG.z.b.V 172. This included Condor S1, D-ADHR *Saarland* – DLH's two other operating Fw 200s, D-ARHW *Friesland* and D-ABOD *Kurmark* remained with the airline for the time being.

KG.z.b.V 172 was formed under the command of Hauptmann Otto Brauer for the transport of troops, supplies and equipment to the front during the invasion of Poland. By the middle of September, German victory was virtually assured, and by the 20th some of the commandeered DLH aircraft could be released back to airline service, and all had returned by October. In reality, this was a temporary reprieve, and most of the DLH Condors would transfer to military service before long.

Nevertheless, during what became known as the 'phoney war' the possibility was raised that previously ordered Condor airliners might be delivered after all. In December 1939, the Swiss

magazine *Inter Avia* reported that two of the five Japanese Fw 200 Bs were to be delivered. This caused a minor flap in British intelligence circles, where it was considered to be an error, but in January 1940, the German Air Attaché to Tokyo gave an interview to the press in which he indicated that he was working to establish an air service "from Manchoukuo across Siberia to Berlin" and that "It was proposed to use a Fokker-Wolf [sic] Condor aeroplane with four engines to carry 24 passengers". A dispatch from the Acting Consul General the following month referred to a visit to Göring by the industrialist Yoshisuke Aikawa, which was "Believed to be mainly concerned with deliveries of machines already ordered", and a marginal note suggests, "Possibly the 5 Focke-Wulf 'Condors' ordered by DNKK in 1939."

In March, *Inter Avia* appeared to confirm that the order for all five Fw 200s had been renewed, in connection with the Japan–China–Thailand service, and on the 28th a British source was "informed by managing director Japan's air transport company today firm is being kept to purchase 5 Condor aircraft by Germany", but that "he does not know what he will now do with these aircraft which were originally intended for European routes". Routes which, needless to say, were rather less practical and attractive in time of war. An intercepted communication from Berlin to Dai Nippon Kōkū Kabushiki Kaisha (Imperial Japanese Airways) Tokyo stated, "FOCKE-WULF willing signing revised contract."

It was apparent that Germany was now trying to hold the Japanese airline to the contract, but the aircraft were no longer required. Intriguingly, one intelligence note refers to seven Condors rather than the five ordered. It is known that discussions took place between Focke-Wulf and a Japanese delegation in late 1938 regarding a military Condor. The Japanese Navy and Mitsubishi expressed interest in a maritime reconnaissance

Hitler's personal Fw 200, A-0 W. Nr. 3099, during its proving trials in an interim colour scheme.

variant armed with three 7.9mm machine guns for licence production.[20] This interest does not appear to have been followed with firm orders, probably due to Japan's preference for working with Junkers. Nevertheless, it may have helped cement the idea of a military variant and begun the work needed to turn the airliner into a reconnaissance-bomber. The mention of seven Fw 200s is intriguing, but probably an error.

The Fw 200 V10, which was converted to the first armed version between August and November 1939, is sometimes described as having been intended for the Imperial Japanese Navy. In fact, as noted, it was to have been an airliner for DLH, and was adapted for the Luftwaffe, specifically for Oberstleutnant Theodor Rowehl's Fliegerstaffel z.b.V.[21]

However, some of the military adaptations probably stemmed from preliminary design work on a prospective Japanese military version.

This, the first true military Fw 200, differed in numerous respects from all the combat variants that would follow. It had a short ventral tub with no provision for bomb carriage but a gun position at the forward and aft end. There was also a single dorsal turret, of a low-profile bubble type similar to the D30, located slightly forward of the wing trailing edge. It was intended as a photo-reconnaissance aircraft and on that basis was fitted with two Rb50/30 cameras. After testing it was transferred to Fliegerstaffel Rowehl and received the code BS+AF.

The V10's career with this unit was short-lived. On 23 November it was due to undertake its first operational flight, a reconnaissance of Scapa Flow. On take-off from Jever, however, the two starboard

20   See Jerry Scutts, *The Focke-Wulf Fw 200 Condor*, Crécy 2010, pp. 37–40.
21   The V10 is listed in the RLM variant table for the Fw 200 C as '*Ausführung "Rowehl"*', or '"Rowehl" Version' – FW 200 Aircraft, Air Ministry, Directorate of Intelligence and related bodies, Intelligence reports and papers, AIR 40/154

engines lost power (in an echo of the loss of the V1) and the V10 crashed off the end of the runway and was written off, though Flugkapitän Martin Koenig and his crew escaped serious harm.

VIP TRANSPORT — FLIEGERSTAFFEL DES FÜHRERS

One of the earliest and most notable military uses of the Fw 200 was with the Fliegerstaffel des Führers (FdF or the Leaders' Flying Squadron), the unit formed to transport high-ranking Nazi officers and politicians, and visitors of high importance. The unit had its roots in an arm of Deutsche Luft Hansa formed under DLH Flugkapitän Hans Baur in 1934, initially with Junkers Ju 52s. It became an official government transport flight, the Flugbereitschaft RLM (Air Ministry Standby Flight), the following year, but was colloquially referred to as the Regierungsstaffel (government squadron). The flight was based at Berlin-Tempelhof.

The Regierungsstaffel was redesignated Die Fliegerstaffel des Führers in September 1939. Many of the senior personnel of this unit were drawn from DLH and were given SS ranks. Other staff were from DLH, retaining their civilian status. On the outbreak of war, Luftwaffe personnel — carefully chosen and vetted — began to be assigned to the FdF. The FdF itself, however, was never part of the Luftwaffe, and fell under the direct authority of the Reich's Chancellery.

By 1938, Baur considered that the Ju 52 was becoming obsolete, its low performance and dated design making it unbecoming of the German leadership. The new Fw 200 was the obvious choice to transport the most important members of the regime, as the symbol of Germany's advancement in civil aviation and with comfort and performance to match or exceed the best airliners that France, the US or the UK could offer. In January 1939, therefore, the eighth and ninth series production Fw 200s were earmarked for the Regierungsstaffel. These were the aircraft S8 and S9, the final pair of A-0 aircraft, W. Nrs.

3098 and 3099, ordered in early 1938, which had been intended for DLH. They had been allocated the registrations D-ACVH and D-ARHU and the names *Grenzmark* and *Ostmark*.

The first of these, 3098, would, initially, retain its identity and its airliner specification. This included capacity for 26 passengers, with three flying personnel and one steward. This aircraft was designated the 'escort' (*Führerbegleitflugzeug*) to Hitler's transport. The second Condor was to be Hitler's new personal aircraft and would receive the designation V3 indicating its unique and, initially, experimental status. This machine inherited the identity D-2600 from Hitler's personal Ju 52, with its unique call sign, and was renamed *Immelman III*, after the First World War ace Max Immelman. (The first two 'Immelmans' had been Ju 52s).

The adoption of 3099 as Hitler's personal aircraft meant significant revisions to the machine from DLH standard, including a custom interior. The cabin was divided into two by a bulkhead as usual, but there the aircraft diverged. Aft of the bulkhead, the layout was similar to that of the passenger aircraft with seating for eleven. The forward compartment was the Führer's personal accommodation, with a large sofa along the left-hand side of the cabin, and on the right-hand side a table with an armchair for Hitler's use.

A mock-up fuselage was constructed to assess and perfect the proposed modifications, which included a special escape system for the Führer. In an emergency, a hatch beneath Hitler's seat could be opened. A parachute was installed in the seat, which would open automatically with a static line attached to the airframe. It simply required Hitler to be strapped into the seat, for the table to be folded up and locked, and the hatch to be released.

The layout had evidently been finalised by 20 April that year, as a scale model of the aircraft with accurate interior (displayed via a removable roof panel) was presented to Hitler by the Regierungsstaffel on his 50th birthday.

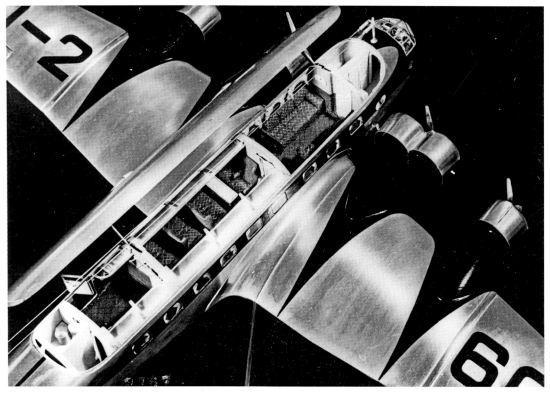

The model of Hitler's personal Fw 200 that was presented to him by pilot Hans Baur on Hitler's 50th birthday, showing an accurate representation of the interior.

As a result, while both Fw 200s were assigned to the FdF at the same time, D-ACVH *Grenzmark* was delivered in June 1939, while D-2600 did not arrive until October, after the outbreak of war. This led to Hitler using *Grenzmark* on occasion until *Immelman III* was available. Both aircraft received colour schemes identical to the DLH Fw 200s, chiefly silver with a black nose sweeping back into a black band along the windows, and black engine nacelles tapering to points on the wing trailing edge.

A prominent early employment for *Grenzmark* was carrying German Foreign Minister Joachim von Ribbentrop to Moscow on 22 August to meet Stalin, formalising the treaty negotiated with Von Ribbentrop's Soviet counterpart Vyacheslav Mikhaylovich Molotov. The Treaty of Non-Aggression between Germany and the Soviet Union,

known colloquially as the Molotov–Ribbentrop Pact, opened the way for the invasion of Poland by both countries, and ended hopes of a Soviet alliance with France and the United Kingdom. Von Ribbentrop returned to Moscow in *Grenzmark* for a further round of talks on 26 September.

On the outbreak of war, *Grenzmark* had a Balkenkreuz added to its registration in place of the dash to create an interim military identity D+ACVH. In this form, it transported Hitler to visit troops at the front just after the invasion of Poland, and then to victory parades in Danzig (Gdansk) on 19 September, and Warsaw on 29 September (D-AETA *Westfalen* may have taken part in some of these flights too). The serial was adjusted to WL+ACVH and then AC+VH, when it was seen in during Hitler's visit to Warsaw-Okecie airport on 5 October 1939 to inspect troops. Soon after this,

Hitler's Fw 200, now in its proper peacetime colour scheme and named Immelman III, just after the outbreak of war when a military Balkenkreuz had been added and the code WL.

the definitive military code NK+NM was allocated.

Hitler's personal aircraft arrived at FdF on 19 October, and Hitler first boarded it on 10 November. During test flights, D-2600 wore a mostly plain silver colour scheme with civilian red tail band and Hakenkreuz in a white disc on the tail, and a black tip to the nose cone. When delivered to the FdF it wore the same scheme as DLH Condors apart from Balkenkreuz markings, with the identity WL+2600. This aircraft too was painted in military colours by April 1940, with the codes 26+00.

The Fw 200 26+00 was the subject of elaborate security measures. It was stored in a secure hangar at Berlin-Tempelhof airport under a standing guard, and everyone who worked on it was carefully vetted. Baur was the only pilot authorised to pilot the Condor and it was claimed that he never revealed the destination of a flight to anyone on the ground.[22]

Both FdF Condors had probably received military colours by the time of the German invasion of neutral Norway and Demmark on 9 April 1940. The colours applied to all Luftwaffe Condors at the time consisted of the standard scheme for daytime-operating aircraft other than fighters: an upper splinter camouflage of RLM 70 *Schwartzgrün* and RLM 71 *Dunkelgrün*, with RLM 76 *Hellgrau* on under-surfaces. The two aircraft evidently received some modifications at this time, and both can be identified by a small air scoop on the nose only seen on these two machines, and only after they received military colours.

In the autumn of 1940, Fw 200 26+00 became the subject of an intriguing plot. On 4 September, the British Attaché in Sofia, Bulgaria, sent a signal to the UK government about an approach he had received from a man calling himself 'Kiroff'.

---

22   R. Moorhouse, *Killing Hitler: The Plots, the Assassins, and the Dictator Who Cheated Death*, Bantam Books, 2007, p. 203.

Kiroff claimed that he was Hans Baur's father-in-law and that Baur, having become disgruntled with his work at the FdF, proposed to kidnap Hitler during a flight and transport him to the UK. The proposal was forwarded to Air Marshal Arthur Harris, then Deputy Chief of the Air Staff, who proclaimed it "fantastic" but put measures in place to take advantage if the story turned out to be true. Security at Lympne airfield was beefed up, measures to ensure secrecy put in place and a plan hatched to transport Hitler to London in a convoy of armed vehicles. The date of the flight was set for 25 March 1941.[23]

Needless to say, 26+00 did not land in Kent carrying the Führer. It's unclear who Kiroff was (he almost certainly was not Baur's father-in-law) or what his motivation was for approaching the British, but it seems highly unlikely that Baur would turn traitor. The pilot remained steadfast to Hitler to the end and stayed in Berlin long enough to be captured by the Soviets when he had had ample opportunity to escape. One possibility is that elements in Bulgaria wished to discredit Baur, who had considerable influence with King Boris.

---

23   Adolf Hitler: Proposed abduction to England by his personal pilot, Air Ministry: Fighter Command: Registered Files, AIR 16/619.

2

1940

THE INVASION OF NORWAY

Despite the relative inactivity of the Phoney War dominating the Western Front in late 1939 and early 1940, moves were afoot to open a new front. Both Germany and the Allies looked with increasing anxiety to north Norway and particularly the flow of Swedish iron ore through its ports. This resource was considered vital to the German war machine, and in the first months of 1940, both sides became increasingly unwilling to let Norwegian neutrality hinder their strategic goals. The British believed that stopping the flow of Swedish iron ore "might well bring German industry to a standstill and would in any case have a profound effect on the duration of the war", considering that controlling Swedish iron ore exports could "bring Germany to the point of surrender".[24]

So important was the matter considered to be that the government started to draw up plans for a partial occupation of Norway, although it was decided early on that such a move could not be made without Norwegian approval. Mining Norwegian territorial waters to drive shipping into international sea where it could be legally attacked was, however, carried out without permission of the government, and secretly the British hoped to provoke Germany so Norway would ask for aid.

Hitler, meanwhile, was increasingly convinced that the British would act unilaterally to gain control of vital resources and ordered that a full-scale invasion of both Norway and Denmark be planned. The result was Operation *Weserübung*, launched on 9 April 1940. This marked the full combat debut of the Fw 200, in dual roles—as a military transport, and in its definitive role as a maritime reconnaissance bomber. Indeed, the evolution of the Condor into the long-range maritime aircraft of legend was already underway.

On the brink of war, Germany's maritime aviation capability was inadequate against powerful naval nations such as Britain and France, in terms of hardware, personnel skill and experience, and organisational structures and processes. While progress had been made in the 1920s and 1930s across all those areas, much of the good work had been lost due to Reichsmarschall Göring's influence on rearmament, the Luftwaffe and ultimately the operation of the entire German military. The Kriegsmarine's influence over maritime and naval aviation was continually reduced, along with the number of units reserved for naval support, until the much-depleted naval air arm was given up

---

24   C. L. N Newall, D. Pound and E. Ironside, Military Implications Of A Policy Aimed At Stopping The Export Of Swedish Iron Ore To Germany, Report by Chiefs of Staff Committee, War Cabinet, 31 December 1939, CAB 66/4/29.

The second production Fw 200 A-0, the DDL airliner OY-DAM *Dania*. This aircraft was in Britain when Germany invaded Denmark and was impounded, being allocated to BOAC.

Scenes at Berlin-Tempelhof after the outbreak of war. In the first image, F8+DU, the former D-AMHC *Nordmark* after joining KG 40, with a communications Junkers W 34, and in the second, camouflaged civilian Fw 200s on the apron outside Tempelhof's distinctive terminal building, with a DLH Ju 86.

The airlift operations for the invasion of Norway in April 1940. Fw 200s of K.Gr. z.b.V. 107 can be seen with Ju 52s and a Ju 90 of the same unit. K.Gr. z.b.V. 107's Condors were camouflaged, while its Ju 90s retained their silver DLH colours.

entirely in April 1942. At the outbreak of war, the Luftwaffe was required to provide aircraft for naval assistance only in the case of actual operations or exercises, meaning there was little opportunity for joint training, and Luftwaffe priorities generally came first. This led to particular problems with navigation over the sea, which made cooperation with the Kriegsmarine particularly difficult.

The Luftwaffe command was increasingly aware of its deficiencies in any maritime war against the UK, and also sought to build up its maritime combat capability for political reasons, taking tasks away from the naval air arm and justifying its claims for control over all forms of military aviation.

In August 1939, X. Flieger-Division[25] was set up under Generalleutnant Hans Geisler as the embryo of a specialist anti-shipping force which would absorb all but a few maritime units from the naval air arm. The Kriegsmarine was left with *Küstenflieger* (coastal) units mostly equipped with seaplanes. One thing the Luftwaffe was completely lacking in was a long-range maritime reconnaissance aircraft with good performance. Hauptmann Edgar Petersen was on temporary detachment to the *Flieger-Division* under Geisler at the time.

Interrogation of prisoners of war who flew maritime reconnaissance missions with the unit Petersen would soon be charged with setting up, described how its creation was

entirely due to the energy of the CO, Major Petersen, who has, for years, made a study of long-range attacks on shipping, and has repeatedly written Minutes on the subject. One of these he sent to General Jenschonnek [sic], who turned down the idea ... However, quite by chance, a

subsequent one was seen by General Kesselring, who became interested in the idea and forwarded the Minute to the Führer himself, which resulted in Petersen being summoned to Obersalzberg for a personal discussion with Adolf. It seems probable that, as a result of this meeting, there will be an extension of this type of activity.[26]

Thus empowered, Petersen had to find an aircraft suitable for the role. It was a short search. On 5 September, Petersen visited the Bremen factory where he would have seen the V10 then under conversion to armed reconnaissance specification for Fliegerstaffel Rowehl, and other Fw 200 B airframes under construction. The FWF engineers reportedly informed him that airframes could be adapted to his specifications in as little as eight weeks from receiving the go-ahead.[27]

Petersen is quoted as later describing the process for selecting the Fw 200:

I looked around for a suitable aeroplane in Germany. There was the Ju 90 but there were only two of these available and no production line had been set up. On the other hand, the Focke-Wulf company had four Fw 200s nearing completion intended for delivery to Japan. I took these and a further six standard Fw 200 transports and with these I set up my Fernaufklärungsstaffel [remote reconnaissance squadron] at Bremen on 1 October 1939.[28]

Petersen's recollections do not entirely line up with the known whereabouts of the existing and under-construction Condors but are close enough to be plausible. The two modified reconnaissance

25 Later X. Fliegerkorps.
26 A.I.I.(k) Report No. 20/1941, Further Report on FW Condor of KG40, brought down 200 miles NW of Ireland, on 10/1/1941, Intelligence reports and papers, AIR 40/154.
27 История самолета #3 (*History of the Aircraft #3*) – Fw 200 Condor, 2006, pp. 38–39.
28 C. Goss, *Fw 200 – The Condor At War 1939–45*, Crécy 2017, p.12.

A Fw 200 with four Ju 90s of K.Gr. z.b.V. 107 in the background during the Norwegian campaign of 1940. The Fw 200 wears a KG 40 emblem on its nose and is therefore most likely F8+HH (the former *Nordmark*) during its spell with K.Gr. z.b.V. 107 from 4 May 1940.

A-0 GF+GF, the former *Saarland* with K.Gr. z.b.V. 107 for transport operations in support of the invasion of Norway, March–May 1940.

aircraft, the V2 and S3, were transferred from Fliegerstaffel Rowehl in November–December 1939. None of the Japanese-ordered aircraft mentioned by Petersen (of which there were five, not four – W. Nrs. 0017–0021) were delivered in their modified C-2 form until August 1940 and then only the first two – the remaining three from the Japanese order were delivered as D-model transports due to the urgent need for transport aircraft during the Norwegian campaign (see below). Other B-model Condors from DLH orders began to be modified right away, however – W. Nr. 0002 was redesignated V11, the prototype Fw 200 C-1 reconnaissance bomber, and W. Nrs. 0003–0008 and 0011–0014 were assigned as C-1 production aircraft.[29]

The C-1 was a rapid modification of the B-model airliner, with the addition of six defensive gun positions (a waist gun on either beam, and two each above and below the fuselage, all rifle-calibre MG 15s apart from a 20mm MG/FF cannon in the lower-forward 'D stand' position). A 'tub' was added beneath the fuselage, offset to starboard, to accommodate bombs, plus bomb racks were fitted on the outer wings (and in later C-1/2 models, an additional bomb rack in a recess in the outer engine nacelle). Six additional fuel tanks were installed in the fuselage to add to the eight tanks in the inner wing between the spars. A small amount of 8mm armour was added to protect the pilot and fuel. There was little additional strengthening despite the weight rising from 17,250kg to a possible maximum take-off weight of 22,700kg. Maximum speed was 378km/h (235mph) and range was 4,500 kilometres (2,800 miles).

The V11 was extensively tested at the E-Stelle,

---

29   Fw 200C Data Table, 3.0 Archiv Werk Flughafen, FW 200 Aircraft, Air Ministry, Directorate of Intelligence and related bodies, Intelligence reports and papers, AIR 40/154.

Fw 200 D-1 W. Nr. 0010 VB+UA wearing its K.Gr. z.b.V. 107 markings including the unit emblem of a steam train, after its conversion to a long-range transport for the Norwegian campaign. Note the twin main wheels.

Rechlin, for the next four months, eventually being transferred to Petersen's unit in April 1940. The squadron became the first *Staffel* of Kampfgeschwader 40 (1./KG 40) on 1 November 1939. Its aircraft were applied with an emblem of the world surrounded by a ring.

Initially, Petersen's Fernaufklärungsstaffel was given the cover identity Kurierstaffel (Courier Squadron) and this appears to have led to the military variant being named, officially or otherwise, the Kurier. An article in *Inter Avia* in early 1941 makes clear the military credentials of the Fw 200 C but uses the name 'Kurier' to distinguish it from the civil version, noting, "The airframe of the Focke-Wulf long-range bomber which now becomes known under the pacific name of 'Kurier' hardly seems to be any different from that of the 'Condor'." The article went on to sum up the

aircraft's armament rather accurately, and propulsion.[30] Similar reports were published in the Osaka *Mainichi* newspaper the following month.

Operation *Weserübung*, the invasion and occupation of Norway and Denmark, was set for 9 April 1940. With another major military operation underway, and one which would require intensive air supply, transport Fw 200s were again sought after. A series of 'special duty combat groups' were formed with the numbers 101–108 and placed under the command of the Chief of Air Transport (Land).[31] In March 1940, unarmed Condors were mobilised, to be assembled in Kampfgeschwader zur besonderen Verwendung 107 (K.Gr. z.b.V. 107), which was made up of large, four-engined airliners under Carl-August von Gablenz. The well-travelled V2 was temporarily transferred from the embryonic 1./KG 40 maritime squadron (see above) to

30   Germany: Focke-Wulf 'Kurier', *Inter Avia* No. 746, 21 January 1941, p. 10
31   Generalmajor a. D. Fritz Morzik, German Air Force Airlift Operations, USAF Historical Studies No. 167, USAF Historical Division 1961, p.92

A Fw 200 C model under construction at FWF's Bremen factory – the under-fuselage 'tub' that marked out the dedicated military versions can be seen, while an engine is manoeuvred to its nacelle.

A Fw 200 of K.Gr. z.b.V. 107 flying low over Narvik during the resupply operation for encircled German forces there in 1940, possibly taken by Unteroffizer Franz Ruckes who was involved in the battles around Narvik with Marine Regiment Berger and who owned this photograph.

K.Gr. z.b.V. 107 from its formation, coded GF+GC (while still wearing the KG 40 emblem), along with the S1, which received the codes GF+GF. These aircraft were joined three days before the invasion by S3 *Nordmark*, also from 1./KG 40 such was the urgency of the transport operation, and this aircraft continued to wear the codes of its earlier unit, F8+HH. After the campaign was underway, K.Gr. z.b.V. 107 was bolstered by DLH aircraft S5 *Friesland*, as CB+FA, and S10 *Kurmark*, as CB+FB. Joining the Condors were Junkers Ju 90s and a solitary Junkers G 38.

Numerous members of the aircrew for this unit and its counterparts were drawn from DLH, with their experience of long-distance navigation at a premium, not to mention their familiarity with the aircraft types they were required to operate. Other personnel were transferred from flight training schools, particularly flying instructors. The mix of existing experience and high levels of basic skill from the instructional crews meant that the new transport groups could very quickly adapt to the tasks they had been assigned to without lengthy specialised training.[32]

On 9 April, the day of the invasion, F8+HH *Nordmark* transported soldiers of Infanterieregiment 324 to the recently captured Oslo-Fornebu airport flown by Oberleutnant K. Verlohr. According to General der Flieger Kessler, "K.Gr. z.b.V. 102 and 107, however, accomplished their tasks at

32   Ibid. p. 93.

A view over the engines of a Fw 200 A investigating a large merchant ship off the coast of Norway in 1940.

Unsere Luftwaffe. Fernkampfflugzeug Focke-Wulf Fw 200 C „Condor"

One of the first publicly released images of a military Fw 200 C-1, W. Nr. 0005 BS+AJ. The censor has removed the A- and B-stand upper gun positions, but the picture shows the under-fuselage 'tub'.

schedule 0830 hours 9 April by landing about 2,000 infantry men within two hours, at first under the fire of the defence at Fornebu. The defence caused some casualties, inter alia a group captain was killed, and the landing was accomplished only at the cost of 31 crashes."[33] Despite these challenges, the airlift was able to proceed as planned on the afternoon of the 9th.

While the south of Norway was overcome within a few days, the north of the country proved tougher, and a strong Norwegian counterattack with support from British and French forces surrounded the German force occupying Narvik. On 13 April, Verlohr took F8+HH to Narvik to airdrop weapons and medical supplies to the besieged troops. Numerous similar supply missions were flown by K.Gr. z.b.V. 107 over the

next few weeks, as maintaining the force in Narvik was a high priority. This became even more so when trains carrying supplies from Sweden under the Red Cross flag were stopped, and aircraft became the only means of support. Initially, only the Condors and Ju 90s of K.Gr. z.b.V. 107 could reach Narvik, and according to a senior officer, "The very sight of these planes which made their appearance from the first days of the invasion whenever weather conditions allowed helped much to keep the troops at Narvik in good morale."[34]

The first of K.Gr. z.b.V. 107's Condors to be lost was not due to combat operations but to a flying accident. Oberleutnant Henke, formerly of DLH, the pilot who had flown D-ACON to New York, overstressed the structure of the S10 CB+FB

---

33 Gen. der Flieger U. O. E. Kessler, The Role of The Luftwaffe and The Campaign In Norway, in D. Isby (editor), *The Luftwaffe and the War at Sea*, Greenhill Books, London Naval Institute Press, Annapolis, Maryland, p. 187.
34 Ibid. p..182.

The first Fw 200 C model, W. Nr. 0002, being refuelled in the snow at Norway alongside Messerschmitt Bf 110 fighters in 1940. The censor has edited out the A-stand. EDWIN HOOGSCHAGEN

Kurmark while manoeuvring near Staaken on 22 April, and a wing broke off. All the crew lost their lives. It was a salutary lesson about the aircraft's delicate structure, designed for gentle passenger flights, not the rigours of war. A second aircraft, GF+GC, the V2, was lost on 28 May when it crashed on the runway at Oslo-Gardermoen and was written off.

Two days after the invasion began, the two Finnish-ordered Fw 200 Bs, then nearing completion at Bremen, were appropriated for the war effort by the Luftwaffe. Like the first two Japanese aircraft, these were to have been adapted to the C model but the urgency for air transport led to their completion as quickly as possible as D-model transports, as were the third, fourth and fifth Japanese Condors. As there were numerous differences

between the aircraft ordered by the two countries, not least that the Finnish machines were powered by Pratt & Whitney Hornet engines rather than BMWs, the former Finnish Condors were designated D-1, while the three machines from the cancelled Japanese order were designated D-2. In May, the Luftwaffe requested that glider-towing equipment be fitted to the Fw 200 Ds then under modification.

With most Norwegian ports successfully captured within 48 hours, the need for air transport diminished and by 8 May, K.Gr. z.b.V. 107 was one of only two of the special transport groups still active in Norway. The unit's Fw 200s were now vital for the resupply of forces in Narvik, as only they and the Ju 90s had the necessary range[35] (although soon they were supplemented with Ju

---

35   Ibid. p. 104.

An early-spec C-1, probably W. Nr. 0005, showing the original form of outer engine nacelles before they were modified to incorporate a bomb rack. The texture of the aircraft's skin is particularly apparent on this image. A fairing for a GV219d bombsight in Ikaria mount, or Lofte 7c in auxiliary Ikaria mount, is fitted beneath the D-stand.

52s fitted with auxiliary fuel tanks). Landing at Narvik was impossible other than for seaplanes, so supplies had to be dropped by parachute or simply jettisoned over the German positions. K.Gr. z.b.V. 107 also brought reinforcements from mountain infantry to parachute in, which was a success despite the soldiers' complete lack of formal parachute training.

Petersen's 1./KG 40 was also engaged to support the beleaguered forces at Narvik by using the Fw 200's prodigious range to harass the Royal Navy and shipping supporting the Allies. The Kriegsmarine had been pummelled by the RN at the first and second naval battles of Narvik on 10 and 13 April, leaving German forces with no naval strength there and suffering a significant blow to

morale. Petersen confirmed to General Jeschonnek that his aircraft were capable of making the flight to Narvik directly from Germany. According to Petersen, a 17-hour flight by 1./KG 40 aircraft was staged, and British warships at Harstad were bombed without visible effect.[36]

Several attacks on shipping are believed to have been made during April, but records do not appear to have survived. 1./KG 40 moved to Gardermoen, north-east of Oslo and from there began to fly armed reconnaissance missions, mostly in the north of the country. The flights were useful to provide information on the location of Allied shipping, both military and merchant, and on numerous occasions, attacks were made too. The priority though was reconnaissance, and sometimes 1./

36   C. Goss, *Fw 200 – The Condor at War 1939-45*, Crécy 2017, p. 13.

Front view of an early C-1, showing the early unfaired bomb racks outboard of the outer nacelles.

KG 40 was also required to assist with the supply operation.

The first of the unit's Condors to be lost was that of Oberleutnant Backhaus, which disappeared without trace on the same day as Henke crashed *Nordmark*, 22 April. 1./KG 40's handful of Fw 200 C-1s continued to reconnoitre Allied shipping and attacking where appropriate. The unit chalked up its first success on 3 May, damaging a Norwegian freighter with a 250kg bomb. The work was risky though, with AA fire from ships a constant threat. Worse, in the middle of May, the RAF succeeded in establishing a temporary fighter base at Bardufoss, with 263 Squadron's Gloster Gladiators now within range of German aircraft supporting the Narvik garrison. The vulnerability of the Condor to fighter aircraft was immediately apparent. On the morning of 25 May, four days after the squadron had begun operations, Flying Officer Francis Ede encountered two separate Fw 200s (though he misidentified them as Junkers Ju 90s) while patrolling Harstad and carried out attacks on both:

At 09.00 hrs, F/O Ede carried out astern attack against a Ju 90 at 15,000 feet 10 miles north of Harstad. F/O Ede was returning from a standing patrol and had separated from No. 3 of his section ... He got in two short bursts at long range and it was later confirmed that this E.A. [enemy aircraft] landed near Dyro [Dyrøya] Island. At 10.30 hrs, F/O Ede contacted a second Ju 90 at 15,000 feet 10 miles S.E. of Harstad. He approached this aircraft from astern down sun, silencing the rear gunner with his first burst and in four successive attacks put each of four engines out in turn. This E.A. was later found crashed in flames on Finnoen [Finnøya] Island, south of Narvik.[37]

Despite Ede only firing two bursts from long range and seeing the aircraft escape, the squadron believed the first aircraft had force-landed on Dyrøya and he was credited with a kill. In fact, the first Condor had escaped unharmed, as

37   263 Squadron Operations Record Book, Summary of events, Air Ministry, October 1939–November 1940, AIR 27/1547.

Fine study of a KG 40 C-1 undergoing maintenance at Bordeaux-Mérignac. The A-stand characteristic of the C-1 and C-2 models is clearly shown, housing an MG 15 (not visible) on a flexible mount covering the forward arc only.

Ede had indicated in his combat report,[38] which contained some interesting details. Despite the misidentification of type, Ede noted some details confirming that it was a Condor. He noted that it had a "single rudder" (the Ju 90 had twin rudders) as well as wearing "dirty green camouflage" (the Ju 90s in theatre still wore a silver colour scheme). He recorded that "Large bombs were observed to drop from outer engines". The early Fw 200 C-1 did not have bomb racks in the engine nacelle itself as later versions would but did have a bomb rack just outboard of the nacelle. He estimated its speed as "250 to 300 [mph]" and found that the Condor was able to outpace his Gladiator by diving to sea level.

The second encounter took place just under an hour and a half after the first and was more decisive: "Red Section was on defensive patrol of Harstad–Skaanland area," according to Ede's report, "on return from previous combat AA fire observed over Harstad at about 10.25 E/A seen going South 15 miles East of Harstad." Red 1 (Ede) engaged and put the aft dorsal gun out of action with his first burst. White smoke was seen streaming from the engines and "trail of same white smoke from port wing root", indicating that the inner wing fuel tank was on fire. The Condor, the machine of Oberleutnant Schöpke, was seen "Going down towards sea when combat finished" and crashed on Finnøya, around 50 miles away from the scene of the combat.[39] Schöpke, along

38   263 Squadron Combat Reports, F/O H. Francis Ede, May 1940, Air Ministry, AIR 50/103.
39   Ibid.

Many of the earlier military Condors had individual names, usually stars or heavenly bodies. F8+AK of 1./KG 40 was named Wega. EDWIN HOOGSCHAGEN

with Feldwebel Fischer, was taken prisoner, while Oberfeldwebel Messer and Feldwebel Börjesson escaped captivity. Obergefreiter Hartleben was killed.

Two days after Ede shot down the Fw 200, 263 Squadron was joined at Bardufoss by 46 Squadron flying Hurricanes. They were immediately in action flying patrols over Narvik as part of Allied operations to recapture the port, which finally succeeded on 29 May. This made things even riskier for the Condors of 1./KG 40. On the day the Allies retook Narvik, Leutnant Freytag was flying a mission to bomb Tromsø, and his Fw 200 C-1 was spotted (and misidentified as a Junkers) by Pilot Officer N.L. Banks, who reported:

I was climbing North towards TROMSØ and sighted a JU 89 flying due West at 12,000 ft. I made an astern attack from slightly below and enemy aircraft turned into sun. Inside starboard engine put out of action and oil from it covered my windscreen, obscuring my vision. Second attack made but I could not get an accurate aim owing to windscreen covered with thick oil. Third attack made but no ammunition left. Enemy flew off with inside starboard engine leaving trail of smoke.[40]

The squadron's Operations Record Book confirms that the aircraft crashed on Dyrøya was

---

40   46 Squadron Combat Reports, P/O N.L. Banks, May 1940, Air Ministry, AIR 50/20-1.

A KG 40 mechanic standing on the starboard outer engine of an early C-3 during fuelling, showing the improved outer wing bomb carriage, with a faired underwing rack and the recess beneath the outer nacelle.

destroyed due to the "Cherif of LANGHAMN who states that the crew of five were picked up dead". (In fact, there were six crew members, who were indeed all killed).[41]

As the 1./KG 40 crews gained in experience and confidence, their results increased, with several more vessels damaged in May, culminating in the sinking of the armed boarding vessel (converted passenger liner) HMS *Vandyck* on 9 June. The *Vandyck* was on its way to escort transports during the Allied evacuation of Narvik when the aircraft of Oberleutnant Heinrich Schlosser found and bombed it off Andöy. Seven of the crew were killed, the rest were taken prisoner after abandoning ship. The wreck remained afloat but ablaze for several days, before eventually sinking. With the withdrawal of British and French forces,

and the escape of the Norwegian Royal Family, by mid-June 1940 the Norwegian campaign was effectively over.

The first of the two Fw 200 D-1s, W. Nr. 0009, was received by K.Gr. z.b.V. 107 on 14 June, shortly after the Allied withdrawal from Norway. It received its unit markings consisting of the code VB+TZ and steam train emblem at the factory. The second such aircraft, W. Nr. 0010 VB+UA, was similarly painted, but K.Gr. z.b.V. 107 was disbanded on 17 June before the aircraft could be delivered. Both D-1s were therefore allocated to K.Gr. z.b.V. 108, but only 0009 flew with this group, still wearing the old unit's markings.

The following month, these aircraft left Luftwaffe service when they were allocated to DLH, for a freight and passenger service between Dakar and

41   46 Squadron Operations Record Book, Record of events, Air Ministry, 1-31 May 1940, AIR 27/460-18.

In the centre is Heinrich Schlosser who sank the troop transport Vandyck on 10 June 1940, one of the first major successes for KG 40 in the anti-shipping role.

Casablanca. In fact, all available Fw 200 transport variants were assigned to this purpose, including D-2 W. Nr. 0019 *Rheinland* which was earmarked to operate a courier service to the Waffenstill-standskommission (Ceasefire Commission) in Casablanca. The other Condors transferred to DLH for the North Africa service were A-0 models S1 *Saarland*, S3 *Nordmark* and S5 *Friesland*. An additional D-2, W. Nr. 0021, was retained by the Luftwaffe for liaison flights to occupied countries, and it was given the code NA+WN.

DIVERSION

The German military command recognised in 1939 that effective attacks on British shipping would be crucial to any war against the UK. However, planning and carrying out the invasion of Norway diverted attention and resources from the commerce war, and while KG 40 proved the

effectiveness of the Fw 200 in the long-range, anti-shipping role during that campaign, it had yet to be let loose on the main Atlantic supply lines.

The attempts to quickly subdue the UK after the defeat of France led to further diversion from this role. There were two reasons for this: first, the decision to prioritise air-mining of British ports, and second, the desire to utilise the Fw 200's range to contribute to attacks on British industry. The Luftwaffe high command believed that shipping should be a secondary target to industry, and for a short period, this overrode the purpose KG 40 was established for.

At the end of the Norwegian operations, 1./KG 40 moved back to Germany, gaining a second *Staffel*, 2./KG 40, and thus becoming a Gruppe – I./KG 40. A *Geschwaderstab* (headquarters unit) would also be formed in July. The supply of Fw 200 Cs from Bremen was still slow but was now at least

An interesting photograph of a KG 40 C-1, issued to the press for propaganda purposes with various details censored, including the code, fuselage tub and B-stand, although the A-stand and KG 40 emblem have been retained. The caption provided to the media read "England's fright. A long-distance bomber Focke-Wulf Condor on a flight over enemy England. Everywhere smoke is to be seen, a sign that once more the German bombs have hit their mark." Fw 200s never flew over the UK in daylight and the 'smoke' is clearly cloud.

sufficient to allow expansion. In mid-July 1940, the *Gruppe* moved to Marx near the north-western coast of Germany under Luftflotte 2. Here they were employed in support of 9. Fliegerdivision, which was mostly based at Soesterberg in the Netherlands, carrying out magnetic mining missions on east coast ports—the Fw 200, on the other hand, could lay mines as far north and west as the Firth of Forth and Belfast. Mining had been carried out spasmodically since the outbreak of war, with disagreements between the Kriegsmarine and Luftwaffe on the timing and volume of operations.[42]

The first minelaying mission by KG 40 took place on 15 July and within four days, the unit suffered its first loss on these operations with Hauptmann Steszyn's aircraft shot down by AA fire while passing Hartlepool. Two of his crew survived and were captured, the others dying in the crash. After another four days, a second Fw 200 was lost when Hauptmann Zenker was forced to ditch his aircraft in Belfast Lough due to an airlock in the fuel system causing two engines to cut out in the midst of the minelaying run. Remarkably, these two losses had robbed KG 40 of the *Staffelkapitäns*

---

42   Vice Admiral E. Weichold, Survey from the Naval Point of View of the Organization of the German Air Force for Operations over the Sea, 1939—45, ONI Review August 1946

Left and above: Three images of Hitler's 26+00 after it had gained a camouflage colour scheme in 1940. Note the small intake on top of the nose which was a feature of the two initial FDF Condors.

of both 1./ and 2./KG 40 in less than a week.[43]

The Condor was not the most suitable aircraft for minelaying, and the small number of aircraft available made such losses untenable. Petersen pleaded for KG 40 to be released from these operations, which was agreed after only a dozen such flights.

On 5 August, KG 40's Condors began to transfer to the base that it would be most closely associated with, Bordeaux-Mérignac in south-west France. Aircraft also operated from Brest. This would enable them to range wide over the Atlantic in search of shipping. Even then, however, the aircraft would occasionally be diverted to the Luftwaffe's bombing campaign against British industry and infrastructure during the Battle of Britain. Once again, the Condor's range meant it could strike at targets no other aircraft could reach.

Occasional night bombing raids against Liverpool were carried out from August 1940, during which one inexperienced crew bombed Dublin by mistake[44] (probably on 2 or 3 January 1941). On 14 October, a single aircraft was dispatched to bomb the Hillingdon Rolls-Royce shadow factory on the outskirts of Glasgow. The same crew returned on the 21st but poor weather over the target meant they were unable to carry out the attack, while three Condors bombed Glasgow and Falkirk on 29 November.[45]

Interrogation of a captured Fw 200 crew in early 1941 noted that

In the middle of December two Condors, each with a bomb-load of 2,500 kilos, carried out an attack on Glasgow. The first started at 2000 hours, and the other at 2300 hours. On arriving at the target area there were heavy cloud conditions

---

43 Sources vary on whether Zenker was in fact the *Staffelkapitän* of 2./KG 40 or whether this role was held by F. Fliegel, but Zenker's Erkennungs-smarke (ID tag) indicated it to be him. Fliegel became commander of I./KG 40 in December.

44 A.I.1.(k) Report No. 20/1941, Further Report on FW Condor of KG40, brought down 200 miles NW of Ireland, on 10/1/1941, Intelligence reports and papers, AIR 40/154.

45 C. Goss, *Lochaber No More*, *Aeroplane*, May 2021, pp. 30–34.

An officer bearing dispatches disembarks NA+WJ W. Nr. 0017, used as a crew trainer and transport aircraft for KG 40 after it had been superseded in the combat role.

over Glasgow itself, and one of the aircraft spent 55 minutes circling the town trying to locate the harbour. One hit was claimed from a height of 1,000 metres.[46]

ATLANTIC HUNTER

The main purpose of I./KG 40 from the second half of 1940 was that which Petersen had considered for them back in September 1939, preying on the shipping that kept the UK supplied. A modest success was achieved on 18 August with damage to a lone Norwegian merchantman off the Bloody Foreland, quickly followed by the sinking of British merchantman SS *Goathland* with its cargo of iron ore from Sierra Leone, on the 25th.

The Condors, though still extremely few in number, began to chalk up an increasing number of attacks on single ships and convoys, which were at that time often lightly protected. The C-1 and C-2 models in use at that time lacked any kind of bombsight and the technique for bombing ships was therefore restricted to what the Luftwaffe high command (Oberkommando der Luftwaffe, or OKL) described as "Low-level lateral attacks by Condors which until then were the usual and only possible method of attack owing to the equipment of this aircraft".[47]

The Canadian Pacific Line's flagship was requisitioned as a troop transport in November 1939 and had been active in bringing soldiers from

---

46   A.I.1.(k) Report No. 20/1941, Further Report on FW Condor of KG40, brought down 200 miles NW of Ireland, on 10/1/1941, Intelligence reports and papers, AIR 40/154.

47   The Operational Use of the Luftwaffe in the War at Sea 1939–43, OKL, 8th Abteilung, January 1944.

Canada and New Zealand to the UK, and from the UK to South Africa. She was returning from such a voyage to Cape Town when spotted by the Fw 200 of Oberleutnant Bernhard Jope 60 miles west of Donegal, during an armed reconnaissance. Jope attacked obliquely from the port bow and dropped six SC250 bombs, two of which hit the liner, bringing her to a stop and starting uncontrollable fires, although she remained afloat. Jope recorded a 'damaged' claim.

Commander B. G. Scurfield, commander of HMS *Broke*, one of several ships called to assist, described the devastation caused by the bombs when the destroyer arrived the following day: "The *Empress of Britain* was burning and was practically gutted apart from the extreme fore part and right aft, although her sides and funnels were intact," adding "She had been abandoned. I considered it unlikely that boarding would be possible for at least 12 hours." He went on to describe the ship as "derelict".[48]

Nevertheless, *Empress of Britain* was taken in tow, but it seems another Fw 200 may have sealed the liner's fate. Scurfield wrote: "While the tow was being secured a four engined Focke-Wulf Condor appeared momentarily and obviously spotted what was afoot. Soon afterwards a Whitley reported that she had sighted a U-boat submerged steering north, 56 miles to the southward."

The U-boat was probably *U-32* commanded by Hans Jenisch, which despite the presence of RN destroyers *Broke* and *Sardonyx*, was able to work into position to torpedo the liner. The propaganda value to the Luftwaffe of sinking so large and prestigious a ship was exploited to the full, and Jope was awarded the Ritterkreuz (Knight's Cross) for his efforts. The board of inquiry into the loss castigated several of the RN officers involved

for failing to take decisions that may have saved the ship, but the incident thoroughly proved the potent combination of Condors attacking directly and providing reconnaissance for U-boats.

Attacks on single vessels and convoys continued through the autumn of 1940 and into the winter, with several more claims for ships damaged or, occasionally, sunk. The Greek merchantman SS *Victoria* was one example of the latter, being part of Convoy SLS 51 having departed Freetown 9 October with a cargo of sugar. The convoy was only a day away from arrival at Liverpool, in the outer approaches to the North Channel, on 30 October when a Condor attacked with two SC250 bombs, causing the *Victoria* to founder.

The activities of KG 40 in the anti-shipping role, however, had made the unit sufficiently dangerous to merit a response. No fewer than 44 RAF bombers undertook a raid against Bordeaux-Mérignac on 23 November, leading to the destruction of two Condors — a new C-3 and a D-1 intended as a training aircraft "which had recently arrived from Bremen".[49] The raid was followed up on 8 December, 27 December and 4 February 1941, with little result.

In December, KG 40 seized the opportunity to make a larger scale of attack than the previous efforts with single aircraft. Ju 88s returning from a sortie to bomb the hydroelectric plant at Lochaber reported a concentration of merchant ships off Oban — in fact, the assembly point for a convoy to Bombay. Seven Fw 200s of I./KG 40 were assigned to attack, but two went unserviceable and the raid went ahead with five aircraft. They claimed hits on three merchantmen of 10,000–12,000 tons, though in fact only one vessel was sunk, the 6,941-ton SS *Breda*, and two others received slight damage.[50]

---

48   Commander B. G. Scurfield, Report from HMS *Broke* re *Empress of Britain*, 30 October 1940, Board of Enquiry into loss of SS *Empress of Britain*, ADM 178/246.

49   A.I.1.(k) Report No. 20/1941, Further Report on FW Condor of KG40, brought down 200 miles NW of Ireland, on 10/1/1941, Intelligence reports and papers, AIR 40/154.

50   C. Goss, Lochaber No More, *Aeroplane*, May 2021, p. 33.

3

1941

MARITIME RECONNAISSANCE

The first model of maritime reconnaissance Condor was the C-1, followed by the similar C-2. The C-1/2 was effective but there was considerable room for improvement with its structural integrity and performance. Just as 1940 was coming to a close, a significantly improved version started to become available to KG 40. The C-3 model incorporated numerous improvements over the C-1/2 and would form the basis of all future versions. W. Nr. 0025 was modified as the prototype C-3, and as such redesignated V13. Production aircraft were from W. Nr. 0026 to 0094.

Most significantly, the C-3 exchanged the C-1/2's BMW 132 engines for BMW Bramo Fafnir 323 Rs. The 323 had a different evolutionary path from the Hornet-derived 132, having its roots in Siemens' licence production of the Bristol Jupiter, and continued development when BMW took Siemens/Bramo over in 1939. The 323 afforded the Fw 200 rather better speed (410km/h, 255mph compared with 378km/h, 234mph) for the same range (4,500 kilometres, 2,800 miles). The 323 Rs were enclosed in close-fitting cowlings with a smaller intake than on the C-1/2, and the outer engines were toed out slightly[51] to improve stability, and controllability

in an engine-out scenario.

Defensive armament was improved too. The forward-facing flexible gun of the C-1/2 was exchanged for a fully revolving D30 mounting in the B-stand (forward upper) position. The fuselage was strengthened somewhat. Only 10 C-1s and eight C-2s were built — the C-3 would be built in significantly larger numbers, although still low by mass-production standards.

Armed reconnaissance missions continued into the new year of 1941, and two successful attacks had been made by 10 January — SS *Temple Moat* was damaged 500 miles west of Cape Finisterre on the 5th, and MV *Clytoneus* sunk 300 miles WNW of Malin Head on the 8th — when a Condor from the Stab. Staffel was shot down during an armed reconnaissance. Fw 200 C-3 W. Nr. 0035 F8+AB was hit by gunfire from the armed tug *Seaman* and forced to ditch. Three of the crew were killed and the other three taken prisoner. This represented revenge of sorts for the *Seaman*, which had been involved in the unsuccessful attempts to save the *Empress of Britain*. The aircraft's pilot, Oberleutnant Burmeister, was Operation Staff Officer of I./KG 40, and was shot down on his last flight before taking up his post as Liaison Officer with

---

51   B. Wheeler, Condor Aerodynamics, *Aeroplane*, October 2021, p. 28.

The view from the D-stand of a low-flying Fw 200 in company with another Condor as they 'beat up' Kriegsmarine patrol vessels. AUSTRALIAN WAR MEMORIAL

the Befehlshaber der U-Boote (B.d.U.), the U-boat operations HQ.

The interrogation of the three crewmembers rescued, pilot Oberleutnant Burmeister, engineer Oberingenieur Gumpert and gunner Unteroffizier Steinmayer, revealed a fascinating insight into KG 40 operations during this early phase in the Battle of the Atlantic. The interrogation is worthy of quoting at length.

The report indicated that at the time of F8+AB's loss, KG40 consisted of two *Staffeln*,[52] each with six FW Condors, and the Stab having three.

The number of flying personnel at Bordeaux is 12 officers and approximately 80 ORs [other ranks], which is amply sufficient for the present number of aircraft, and scale of effort. Such losses as they have had, have been very quickly replaced.

The Condors normally carry a crew of six, consisting of 1st and 2nd pilot, observer, W/T operator, B/M [engineer] and Gunner. The 1st pilots have all obtained their Teachers' 'C' Certificate, and the 2nd pilots are said to be all experienced men drawn from the z.b.V. Units. The observers all add Astro-navigation to their other qualifications, and in most cases are capable of taking over the W/T.

The crew of this particular aircraft, although six men, was not entirely ortho-dox. The 1st and 2nd pilot, W/T operator

---

52   KG 40 actually gained a third *Staffel* at the end of December and it became active around this time – in fact, an aircraft from 3./KG 40 was reported lost the day after F8+AB was shot down.

Fw 200 C-3/U2 wearing the temporary codes SG+KR – this aircraft would become F8+AH with 1./KG 40 and would be destroyed in February 1941. The censor has removed the A-stand.

and observer, were as usual, but a *Leutnant* went as extra observer because he wanted the trip, and a *Flieger Oberingenieur* who belonged, strictly speaking, to the ground staff, had taken up flying because he wanted to win the Iron Cross. So far he is still without the E.K.II.[53]

The information received in the interrogation, while it could not be absolutely relied upon as accurate, nevertheless gave a detailed and compelling picture of the anti-shipping and other operations carried out by the Condors of I./KG 40. From a historical perspective, much of what was recorded can be corroborated. The picture of the general activity of the unit is valuable for being recorded in real time, as opposed to recalled after the event, and free of the political manoeuvring that often-characterised accounts of Luftwaffe maritime operations either by the Luftwaffe or

the Kriegsmarine. The record continues:

The armed reconnaissance patrols, which form the main function of KG 40, usually start very early in the morning, the men get up at about 3 a.m., the aircraft are away by 5 a.m. and arrive at their Patrol Area at about 0930 hours, while Venus is still visible low down in the heavens. At present patrols are always carried out by aircraft operating singly, but on occasion two aircraft may cross on patrol, or one may be relieving the other in the same patrol area.

The interrogation described

Four main routes used by these aircraft from Bordeaux. These are carefully worked out according to petrol consumption, and are adhered to rigidly, at least on the

---

53   Eisernes Kreuz II. Klasse (Iron Cross 2nd Class)

'Mars', F8+AB of Stab. I./KG 40, being 'bombed up' before a mission, with an SC250 bomb on a trolley.

outward course. The return journey is probably more flexible, dependent upon the activity in the patrol area. No fighter protection is provided, either on the way out, or to cover the return. All these routes entail a round trip of approximately 3,500 kms., and normally the procedure is to fly at a height of 500/700 metres, with the engines throttled back to given an air speed of about 270 kph.

The routes were as follows:

1. Usually skirting the Spanish coast down to Lisbon, then out to latitude 17° or 18° West. From this point the track turned northwards towards Ireland, and then returned to Bordeaux

2. Due West from Bordeaux to between 17° or 18°, and then back to Bordeaux after a long sea patrol

3. Flying close to the coast of Brittany, skirting Cornwall to the Irish Sea and then across Ireland to Ellis Head, before turning north-west for 570km, then back to Bordeaux, via the Irish Sea or sometimes direct. (On this route the aircraft return to Bordeaux though it would shorten the journey significantly to land in Norway, because weather conditions were too difficult in winter.)

4. Mostly far out to sea, crossing only the SW corner of Ireland, or passing just off Mizen Head.

Initially, because of fears of British retaliation if Irish neutrality was violated, the crews had orders not to cross Ireland except in a case of emergency. By early 1941, however, this had been relaxed and "Flight across Ireland at between 1,200/1,500 metres is a frequent occurrence. On these flights they are unmolested by AA or fighters".

C-3 F8+AD of Stab. III./KG 40 running up in a dispersal pen. This aircraft went to the Mediterranean in late 1942 to help transport petrol to the Afrika Korps.

The PoWs noted that

These patrol routes are carefully worked out on a basis of their knowledge of convoy routes. They believe that the Convoys normally use latitude 20° West as a highway; Convoys from Gibraltar sailing westwards to the 20° of latitude, where they meet the South American Convoys. The two Convoys then combine, and sail practically unescorted northwest up to the 20° line, to the point where the North American and Canadian Convoys are met, and a strong escort picked up. From this point the Convoy turns eastwards towards the NW of Scotland, and then southwards again hugging the West Coast of Scotland.

Even the long range of the Fw 200 did not allow the Luftwaffe to take full advantage of this knowledge, however, as "These patrols by KG40 only extend to the 17th° or 18th° latitude which, because some period must be allowed for the patrol, is the usual limit of the Condor's range."

The crew indicated that "The patrols do not normally take them down south of Lisbon, but on one occasion, one of their aircraft claimed to have bombed and sunk a 5,000 ton merchant ship off Cadiz". This could be the aforementioned *Temple Moat*, attacked but not sunk at 44.27N 18.55W, still some way north of Cadiz, but by far the southern-most claim registered until this time.

While the offensive actions of the Condors gathered most of the headlines, it was arguably their reconnaissance activities that were most important.

The aircraft of KG40, although their official title is now 'Fernbomber' or 'Long Range Bombers', also carry out general and

Fw 200 C-3 undergoing engine maintenance in the hangar at Bordeaux-Mérignac.

The tail of F8+AL showing 10 credited sinkings between December 1940 and May 1941, plus 13 operations over 'England'. This aircraft was flown four times by Bernhard Jope, one of the most successful of all Fw 200 pilots.

weather reconnaissance. The most important area for weather reports is towards Iceland, and in no case is a report sent off from the aircraft until the patrol area has been reached.

A weather report is almost invariably transmitted at 1100 hours each day, and subsequently as weather conditions require. These reports are compiled by the ordinary crew of the aircraft, but on special occasions when a particularly accurate report is required, for instance, before the threatened invasion last autumn, a 'Weather Frog' will be taken on the flight replacing the rear-gunner.

Here, an intriguing fragment was added by the PoWs: "Dr STORM, whose body with one other,

An unknown Fw 200 C-3 showing the KG 40 'world in a ring' emblem.

Oberleutnant SCHULT, P/W think was recently washed ashore on the coast of Ireland, was one of the most prominent Weather Frogs." This was the aircraft of Oberleutnant Schuldt, which did indeed have a Dr Hans Sturm among its crew when it was lost on 22 August 1940 – the bodies of the two named washed ashore sometime later.

Another meteorologist, Dr Erhard Herrstörm, was killed when F8+AH, the aircraft of Ober-lteutnant Paul Gömmer crashed into a hill at Cashelfean, County Cork, on 5 February 1941. All but one of the crew were killed. There is some confusion in published sources over the loss of this particular Condor, many stating that it was shot down by a merchantman, some naming it as SS *Major C* or SS *Majorca*, or that it had attacked a Greek merchant ship earlier. In fact, there was no such vessel as the *Major C*, and the only possible SS *Majorca* was nowhere near the scene of the crash at the time. Every indication from the crash is that the Condor flew into the hillside due to poor visibility or a navigational error. As confirmed by the crew of F8+AB, Condors often overflew Ireland while en route to their patrol area, and F8+AH was carrying all its bombs when it crashed, suggesting it had not been in trouble, and had not previously carried out an attack.[54]

At least enemy aircraft were not yet presenting much of a threat. The flying boats which then constituted RAF Coastal Command's long-range strength were both underperforming and under-armed in comparison with the Condor, as confirmed by an engagement between a Fw 200 and a Short Sunderland of 201 Squadron on

54   Condor loss 5 February 1941, Luftwaffe and Allied Air Forces Discussion Forum, July 2010, http://forum.12oclockhigh.net/archive/index. php?t-21843.html.

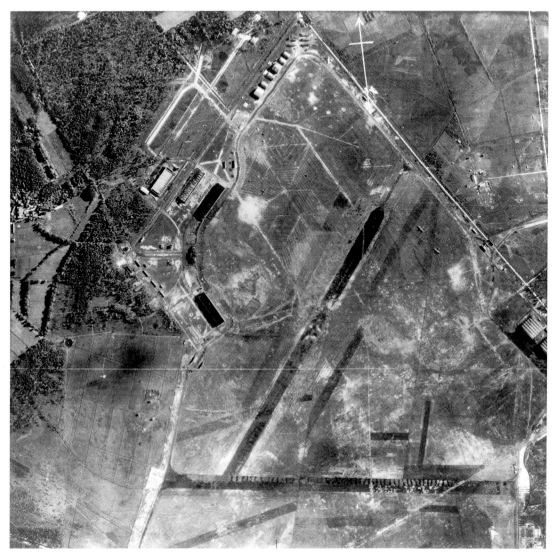

(Above and right) An aerial reconnaissance photograph of KG 40's Bordeaux-Mérignac base, with a close-up highlighting two Fw 200s parked out on the airfield. The 'feather' lines towards the north of the airfield were revealed by prisoner of war interrogations to be drainage channels.

11 January. The squadron recorded:

Condor was on opposite course to that of our own aircraft and 500 ft above. Own aircraft opened fire at 400 yards range and it was estimated that a burst of 8 rounds entered E/A. Unable to get further bursts in owing to inability to close. Our fire was not returned and after circling round our aircraft 3–4 times E/A made off on course 020. A/C could not follow owing to inferior speed.[55]

---

55   201 Squadron Operations Record Book, Record of events, Air Ministry, 1-10 Jan 1941, AIR 27/1178-1.

In fact, the first months of 1941 proved to be startlingly successful for the Condors of I./KG 40. Most famously, an attack on Convoy HG 53 in strength led to the sinking of five merchant ships on a single day. On the evening of 8 February, I./KG 40 were forwarded a report from *U-37* indicating that the 21-ship convoy HG 53 from Gibraltar was in range of its aircraft. Typically, Gibraltar convoys sailed out into the Atlantic as far as 20° west in order to stay out of range, but HG 53 was only at 16° west when reported. Five aircraft took off early the following morning, under the leadership of Hauptmann Fliegel, the flight including Leutnant Jope who had fatally damaged the *Empress of Britain* the previous October.

U-37 helpfully tracked the convoy and sent homing signals for the aircraft. At around noon, after six hours of flight, the Condors sighted the convoy at 35.42N, 14.38W, and immediately went into the attack, dropping as low as possible to skim the wavetops. Each pilot picked a target from the 19 ships still left — *U-37* had sunk two in the early hours of the morning — while the escorts "Engaged with any gun that would bear, whether H.A. [high-angle] or L.A. [low-angle] Gun fire".[56] Despite this, the Condor crews were not deterred, took their time and some were even able to save bombs and attack more than one ship. Oberleutnant Schlosser was credited with sinking no fewer than three, while Hauptmann Fliegel claimed two and Oberleutnant Jope, one, while Leutnant Buchholz was credited with damage to a merchantman. Only Oberleutnant Adam's aircraft failed to inflict any damage on the convoy, having been hit on the run-in by AA fire from HMS *Deptford* and SS *Vanellus*. His Condor limped away trailing thick white smoke due to a punctured fuel tank, and he force-landed in Portugal, all the crew

56   Summary of Report of Ocean Escort — Convoy HG 53 from the Commanding Officer, HMS *Deptford* to Captain "D", Liverpool, 16 February 1941.

Above and previous page: Fw 200 C-3s of KG 40 at Bordeaux-Mérignac, running up before an operation, parked in a dispersal pen between wooden blast panels, and in front of the airfield's distinctive hangars. F8+CH shows the remains of its 'KE' delivery codes beneath its KG 40 identity. Note the mission markings on the tail.

surviving, and managing to partially destroy their aircraft before capture. This aircraft was the second production Fw 200 C-1, W. Nr. 0003.

The true damage to the convoy was four ships lost outright, the *Britannic* sinking immediately, while the *Jura* and *Tejo* sank shortly afterwards, and *Dagmar I* was immobile and had to be abandoned – tugs were sent out from Gibraltar but failed to find her. The *Varna*, meanwhile, was badly damaged and while able to continue, struggled during "mountainous" seas that hit the convoy a few days later and sank on 15 February. The attack by I./KG 40 had left the convoy in disarray, and it took the remaining daylight hours for the escorts to get it back into order. *U-37* had remained in

contact and sank another merchantman, the *Brandenburg*, the following morning.

Further successes would be added before the month was out. Ten days later the indefatigable Leutnant Jope encountered Convoy OB 287 during a flight that was supposed to be a weather reconnaissance and sank the merchantman *Gracia* and tanker *Housatonic*. Not satisfied with that, Jope located stragglers from the same convoy on 21 February and bombed two ships, severely damaging the *Scottish Standard* and facilitating its sinking by *U-96* a few days later.

The report into Oberleutnant Burmeister's crew described the method of attack:

Fw 200 undergoing engine maintenance – on the left is Richard Müller, from engine manufacturer BMW.

Further interrogation has confirmed, that attacks on shipping are carried out from a very low altitude, usually between 70 and 80 metres, using Fuse 38.

From this low altitude the bombs are always dropped Vz, which according to a previously captured document gives a total safety time of 6.2/7 seconds. This table gave the minimum height of release as 20 metres, and in practice KG 40 sometimes bombs ships from masthead height, though this has the disadvantage that the time of fall will be less than 3 seconds, and P/W think that if a near miss is scored, the bomb will have sunk too deep into the water before exploding and will not do any considerable damage.

"the attacks are invariably made along the length of the ship from bow to stern, as they believe that in most cases the AA armament of merchant shipping is mounted on the stern, and is 'blind' forwards. If only our merchant ships were armed with four 2 cm AA guns, two at the bows and two on the stern, they consider that their present tactics become suicidal.

When attacking a convoy, they would pick out the largest ship for attack, or any stragglers. This crew made a specialty of ocean-going tugs because they think they are useful to us for towing damaged ships to harbour. It was by HM Armoured Tug 'Seaman' that this aircraft was shot down.

Of the Fw 200's successes of February 1941, perhaps the greatest did not involve the direct sinking of any ships. The attack on Convoy OB 288 of 23 February has been described as a "Classic Wolf Pack operation" and "a textbook operation of how the Luftwaffe can work in combination with the U-boat fleet to wreak devastation on the convoys".[57]

A Condor of I./KG 40 spotted OB 288, which left Liverpool on 18 February, while proceeding westward across the Atlantic around 300 miles south of Iceland. According to the First Mate of SS *Cape Nelson*, "Convoy was attacked by enemy aircraft. The aircraft followed down 5th column, dropping bombs and machine gunning."[58] Two ships, the *Kingston Hill* and the *Keila* were damaged, and had to be escorted back to the UK.

The Condor then reported the position, which allowed the Kriegsmarine U-boat command to direct every available submarine to the position. Six German and two Italian submarines converged on the convoy, whose escort had detached, and reached it on the 23rd. (In the meantime, SS *Waynegate* reported another "Focke Wolf attack" on the 22nd). The attacks began at 2327 and within 24 hours, 12 ships totalling 62,595 tons had been sunk. One of these was the Norwegian freighter *Svein Jarl*, which had been damaged in one of the earliest attacks by a Condor, back in August 1940. The incident proved how destructive the combination of Fw 200 reconnaissance and homing for U-boats could be. Fortunately for the Allies, this approach was not adopted nearly as systematically as it might have been. It is notable that this series of successes came during a period when KG 40 was under the operational command of the Kriegsmarine, before Reichsmarschall Göring wrested it back under Luftwaffe control.

Coordination between the Kriegsmarine and Luftwaffe was difficult even on a practical level. Each used different map grids, and different code and cypher systems, and there was no common communications network. Added to the ongoing political wrangling between the two services, it is remarkable that any coordination was achieved at all. It was mainly due to a willingness to make

---

57   World War Two Daily, February 23, 1941: OB-288 Convoy Destruction. https://worldwartwodaily.filminspector.com/2017/02/february-23-1941-ob-288-convoy.html.

58   Report by 1st Mate – SS *Cape Nelson*, Convoy OB 288 Misc. reports, Warsailors.com, www.warsailors.com/convoys/ob288reports.html.

Above and left: Engine maintenance on a Fw 200, demonstrating how the clamshell cowlings allowed easy access to the BMWs.

things work at a local command level that enabled practical difficulties to be surmounted.[59] Admiral Dönitz was still frustrated at the level of coordination between Condors and U-boats, which he felt could be improved significantly. "There is at the moment very little effective cooperation between aircraft of KG40 and U-boats," the PoW report from Burmeister's crew stated.

Any message reporting Convoys has to go through two different headquarters, and be decoded and encoded twice, with the result that the information is at least 1½ hours old by the time it reaches its destination. It was no doubt to overcome this, and to work out some system of direct communication between aircraft and U-boat, that the post of Liaison Officer had been created. 'Fuhlungshalter' [contact holder]

procedure between aircraft and U-boat is not at the moment being carried out.

Indeed, while Condors could and did home U-boats onto targets, communication was not directly with the submarines. U-boats had to receive the homing signal and re-transmit it to B.d.U. which plotted the beacon signals and reported the position back to the U-boats. The delay inevitably introduced inaccuracy.

And even when the relationship was at its closest, the Kriegsmarine's commanders often felt that the Luftwaffe was working to its own agenda. A Kriegsmarine Staff Operations officer later wrote that

Instead of attacking destroyer escorts and sloops in the convoys on the North Atlantic or Gibraltar route — leaving the cargo ships

59   See L. Paterson, *Eagles Over the Sea*, Seaforth 2019, p. 137.

Atmospheric press photo of a KG 40 Condor – the contemporary caption labels the aircraft 'Kurier' rather than 'Condor'.

to the U-boats – the Luftwaffe attacked the cargo ships, scattering them and leaving the escorts free to attack the U-boats; but they could report sinking an exaggerated amount of tonnage forgetting that in proper teamwork, the two services could have bagged the whole lot.[60]

Nevertheless, reconnaissance missions were carried out to the benefit of the Kriegsmarine, even while the Luftwaffe tended to prioritise its own objectives. A particularly epic example took place on 27 March, when Bernhard Jope flew the D-2b W. Nr. 0020 F8+GL to survey the state of the sea ice in the Denmark Strait.[61] He took off from Stavanger, passed close to Iceland and reached the

coast of Greenland, bombing the town of Scoresbysund (Ittoqqortoormiit) as demonstration. Jope landed at Trondheim after covering 5,000 kilometres and spending 17 hours aloft.

Between February and March 1941, the number of Condors on strength at I./KG 40 rose from 15 to 21, and in May to 25 to 30, although often a large proportion of these aircraft remained unserviceable. The rate of Condor production by FWF was sufficiently low that when a second and third *Gruppen* were formed within KG 40 in the first half of 1941, they were equipped with Dornier Do 217 and Heinkel He 111 bombers. Also in March, the brief period of Kriegsmarine operational control of I./KG 40 came to an end, although there was a sop thrown to the navy in the creation of

60   Kapitän-zur-See H-J. Reinecke, Cooperation of the Luftwaffe with the German Navy during World War II, in D. Isby (editor), *The Luftwaffe and the War at Sea*, Greenhill Books, London Naval Institute Press, Annapolis, Maryland, p. 221.
61   История самолета #3 (*History of the Aircraft* #3) – Fw 200 Condor, 2006, p. 45.

FW 200 V1 D-AERE

The Fw 200 V1, first prototype *Brandenburg*, as it appeared for the majority of its flying trials in 1937-38 before rebuild with swept outer wing.

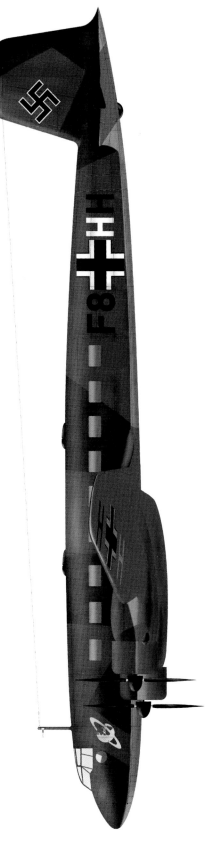

FW 200 A F8+HH

Fw 200 A-0 W. Nr. 2895 *Nordmark*, 1./KG 40 November 1939–April 1940 until joining KG z.b.V. 107 in May 1940 for transport duties in Norway. Camouflage is a splinter pattern of RLM 72/73 over RLM 65.

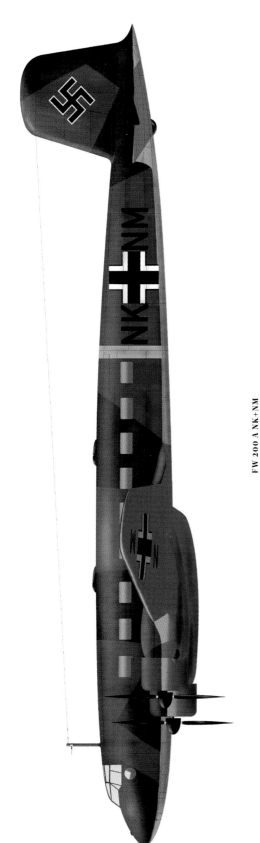

FW 200 A NK+NM

Fw 200 A-0 W. Nr. 3098 *Grenzmark*, Fliegerstaffel des Führers, as it appeared from the summer of 1941 to its loss in December, with yellow Eastern Front identification band.

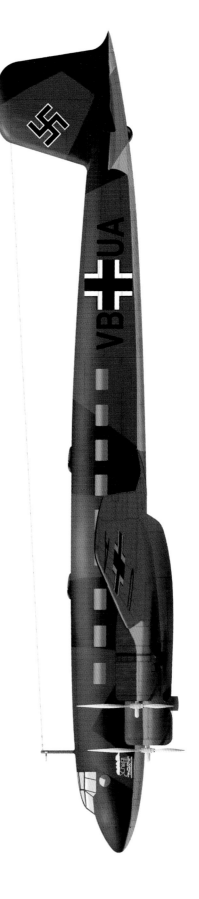

FW 200 D VB+UA

Fw 200 D-1 W. Nr. 0010, VB+UA wearing unit emblem of transport squadron 4./KG z.b.V. 107, June 1940. This aircraft later served with DLH as an airliner and finally as a transport with KG z.b.V. 108.

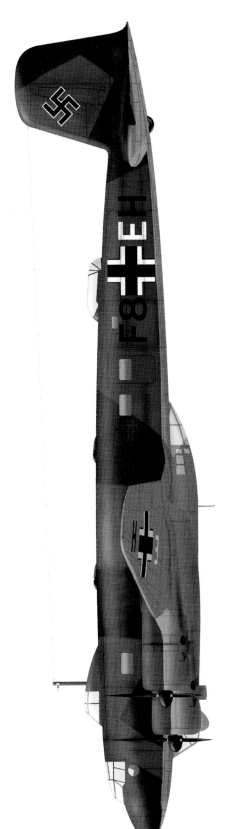

FW 200 C-1 F8+EH

Fw 200 C-2 F8+EH, W. Nr. 0023, possibly named *Elektra*, which arrived at 1./KG 40's Lüneburg, Saxony, base on 31 July 1940 as replacement for the previous F8+EH, which had been shot down on a minelaying sortie.

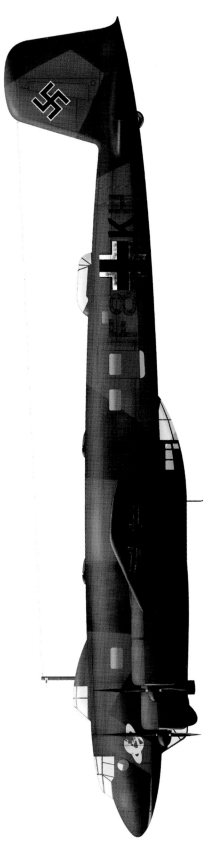

FW 200 C-1 F8+KH

Fw 200 C-2 F8+KH, W. Nr. unknown, of 1./KG 40 with a temporary black camouflage covering the RLM 65 undersides during the *Staffel*'s brief campaign of minelaying and bombing land targets in the UK at night.

FW 200 C-1 F8+DK

Fw 200 C-1, W. Nr. 0003, F8+DK of 2./KG 40, flown by Oberleutnant Adam during an attack on Convoy HG 53, 8 February 1941. This aircraft was hit by AA fire and force-landed in Portugal, being destroyed by its crew. Mission markings on the tail record 10 missions to Narvik and 18 to England. The code had either been toned down or worn off.

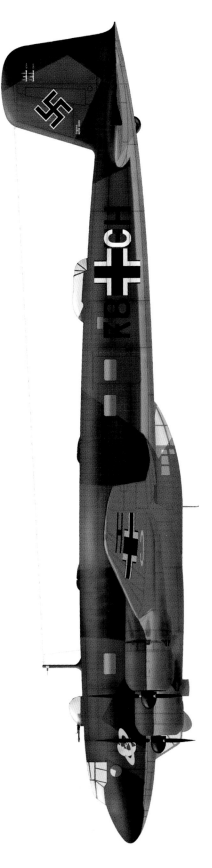

FW 200 C-3 F8+CH

Fw 200 C-3 F8+CH *Capella*, W. Nr. 0026, of 1./KG 40, flown at various times by Hauptmann K. Verlohr, one of the earliest and most successful military Condor pilots. It wears two ship 'kill' markings and numerous bombing mission markings. The delivery codes are clearly visible beneath the unit codes.

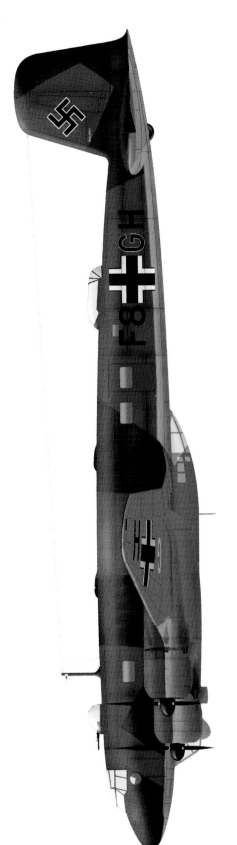

FW 200 C-3 F8+GH

Fw 200 C-3 F8+GH, W. Nr. 0074, which formed part of a detachment to the Mediterranean in August–September 1941, based at Eleusis near Athens, for attacks on Suez shipping. It was lost on 5 September.

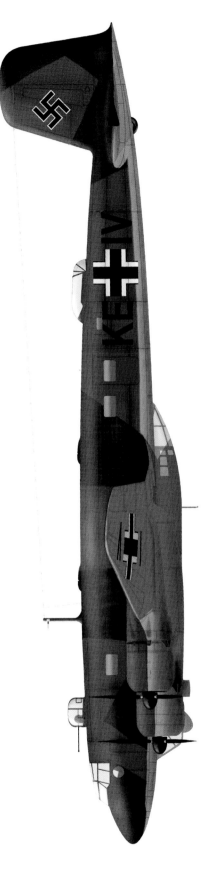

FW 200 C-4 KE+IV

Fw 200 C-4 W. Nr. 0097, in its delivery codes KE+IV. This was a long-serving aircraft with KG 40, serving with 3 Staffel, 8 Staffel, and latterly Transportstaffel 'Condor', with which it was lost in a crash at Nautsi, Finland, 17 October 1944. It was fitted with the large, hydraulic HD 151 A-stand turret.

FW 200 C-3 SG+KS

Fw 200 C-3, W. Nr. 0043, wearing its factory codes SG+KS. This C-3 was built with the new, improved 'lens' D-stand gun position in the forward fuselage tub. It was flown by Hauptmann F. Fliegel, long-serving KG 40 officer, when destroyed by AA on 18 July 1941.

FW 200 C-4 F8+GT

Fw 200 C-4, W. Nr. unknown, from III./KG 40 at Lecce, Italy, providing fuel transport for the Afrika Korps in late October 1942.

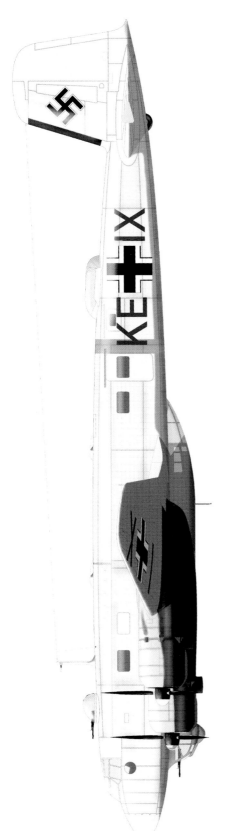

FW 200 C-4-U1 KE+IX

Fw 200 C-4/U1, W. Nr. 0099, Fliegerstaffel des Führers, from January 1942 when it appeared in this temporary white, winter scheme. Like most of the later FdF Condors this aircraft had black-painted engine nacelles, though for the winter scheme the upper parts were covered with white.

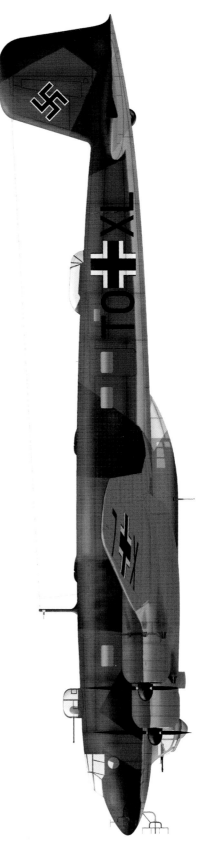

FW 200 C-8 TO+XL

Fw 200 C-8 W. Nr. 0256, first of the second batch of C-8s manufactured from C-5s. This machine lacked some of the features commonly associated with the C-8 (the lengthened fuselage tub and deeper outer engine nacelles) but included the second 540-litre fuselage tub fuel tank. FuG 200 Hohentweil radar was standard for C-8s.

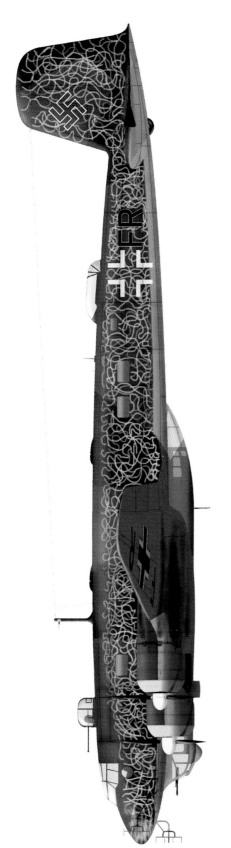

FW 200 C-8_U10 F8+FR

Fw 200 C-8/U10 F8+FR, W. Nr. unknown, 7./KG 40 at Gotenhafen-Hexengrund, May 1944. *Mäandertarnung*, or 'meandering camouflage', consisting of a standard scheme for day aircraft with a *Wellenmuster* or 'wave pattern' in RLM 76 applied over the splinter pattern, for over-sea camouflage. This version has the deeper nacelles for glider bomb carriage.

FW 200 C-8 NO WING

Generic Fw 200 C-8 with wing removed to show fuselage tub. This example has the longer fuselage tub introduced on some C-8 airframes.

C-3/U2 W. Nr. 0042, F8+AH with 1./KG 40, at night. This aircraft would be lost in February 1941 when it crashed into high ground in County Cork. The improved D-stand with 'lens' mount for machine gun (not shown) can be seen clearly.

C-4 flying low over the sea past a Kriegsmarine patrol vessel, clearly showing the larger HD151 A-stand turret. This was hydraulically powered and carried a 15mm heavy machine gun, but increased drag and reduced the aircraft's top speed.

Major Bernhard Jope, one of the most successful exponents of the Fw 200, who sunk numerous ships including fatally damaging the *Empress of Britain*.

A German propaganda postcard published by the magazine *Adler*. The caption reads: "German long-distance bomber Fw 200 over a wrecked British steamer. The crew escaped into the boats."

Fliegerführer Atlantik, a role intended to oversee aerial reconnaissance for B.d.U. and surface forces, and which I./KG 40 would report to. The post was ably filled by Oberstleutnant Martin Harlinghausen, a veteran naval aviator, although Harlinghausen would soon accompany JG 26 to the Mediterranean and therefore be absent from the Fliegerführer Atlantik HQ.

Though it was not apparent at the time, the apogee of the Fw 200's career had already passed. The successes up to that point had been permitted by light or non-existent AA armament on merchant ships, and a paucity of escort vessels. Much work was done to address these shortcomings in 1941, and soon the Condors would find success came harder, and at higher cost. As Oberleutnant Burmeister's crew had identified,

increasing the strength of merchantmen's AA armament made a huge difference to the survivability of attacks, considering the tactics employed. From March, the success rate would start to decline, and the loss rate, increase.

In March, one aircraft was lost, while in April that rose to four. In May, only one was lost but in June the losses went up to four again, although two of those were to an RAF raid on Bordeaux-Mérignac. The experience of Convoy HG 65 in June 1941 was indicative of the changing picture. A series of attacks by five Condors led to damage to seven merchant ships, but three Condors sustained damage during the attack — one force-landed in Spain, and its crew was able to make repairs and return it to France; another was on fire and managed to reach Portugal but crashed and blew

A British propaganda illustration depicting the launch of Lieutenant Everett's successful engagement of a Condor in a Hawker Sea Hurricane launched from the Fighter Catapult Ship *Maplin* in August 1941.

Two photographs of the dispatch of Oberfeldwebel Bleichert's Condor F8+BB, by a Lockheed Hudson of 233 Squadron RAF, on 23 July 1941. Ships of the convoy the Condor was attacking can be seen in the background of the first image.

OUTSTANDING EXPERIENCE WITH NUMEROUS TYPES, BRITISH AND FOREIGN, SOME OF WHICH ARE SEEN HERE, HAS EQUIPPED THIS ORGANISATION TO DESIGN AND CONSTRUCT AIRCRAFT FOR THE FUTURE TO MEET, ECONOMICALLY AND EFFICIENTLY, THE NEEDS OF PASSENGER, PILOT, OPERATOR & MAINTENANCE ENGINEER

CUNLIFFE-OWEN AIRCRAFT LIMITED
ENGLAND

The unusual sight of a British company advertising using a German aircraft in 1941. Cunliffe-Owen refurbished the impounded ex-Danish Condor OY-DAM, seen top left, but sadly it crashed shortly afterwards and was written off.

up with the loss of the entire crew;[62] the third limped back to Bordeaux. Later that month, a modified Fw 200 C-3/U3 with an experimental radar went missing over the Atlantic, its loss confirmed when the body of a crewmember washed ashore on the Irish coast.

During May 1941, I./KG 40 joined the Luftwaffe's ultimately vain attempts to prevent the loss of the battleship *Bismarck*. Long-range patrols aimed to locate the Royal Navy forces hunting the battleship, and after she was mortally wounded by a torpedo from a Fleet Air Arm Swordfish on 26 May, armed reconnaissance flights attempted to bring U-boats into the fight to keep the RN off the *Bismarck*'s back. At 0945 on the 27th, a Condor identified Force H on its way from Gibraltar and sent homing signals to B.d.U. for the next hour, ultimately without effect. The *Bismarck* was already doomed.

Despite the higher loss rate, in the summer of 1941, I./KG 40 was able to send out more aircraft than had previously been possible, with up to eight machines in the air at once, four times more than had been possible at the beginning of the year. The *Gruppe* had shifted from attacks in the North Channel off the coast of Northern Ireland to the seas off Portugal.

Protection for the convoys continued to increase, however. Aware that the Condor was extremely vulnerable to attack by fighters, in May 1941 the Allies introduced a scheme that had been proposed in January 1940, and in serious consideration since December that year — fitting catapults to ships to enable them to launch a single-seat fighter. There were two initial schemes, one to fit hydraulic catapults designed for cruisers to large fleet auxiliaries, and the other to fit a simple

62   Report on German Focke Wolfe Kurrier [sic] 200 Aircraft crashed in Portugal, 9 July 1941, Intelligence reports and papers, AIR 40/154.

A series of images from an air-to-air photoshoot of Fw 200 C-3 W. Nr. 0043 SG+KS during a manufacturer's test flight. The first image is untampered with, while most others demonstrate some censoring. The second shows airbrushing to remove the outer nacelle and outboard bomb racks; the third shows A-stand and B-stand removed; the fourth shows B-stand only removed; the fifth shows codes removed; finally, the sixth appears to show no censoring, though it is captured from an angle that obscures the upper gun positions.

A rare image of FdF Condor NK+NM in colours it only wore for a short period, with yellow rudder and elevators as an identification marking on the Eastern Front, possibly in Greece. This was soon replaced with a yellow band around the rear fuselage.

catapult to merchant ships. The former became the Fighter Catapult Ship (FCS), the latter the Catapult Armed Merchantman (CAM Ship), thanks to the rapid development of the cordite rocket-powered 'P' Catapult. A sled carried a fighter, propelled to take-off speed by a battery of 3-inch solid rockets along a rail on the ship's foredeck, for a single defensive flight against a Condor. Pilots were to be supplied from an RAF unit, the Merchant Ship Fighter Unit (MFSU) but initially some naval pilots were assigned to plug gaps until enough RAF pilots were trained.

Six catapult-equipped ships joined convoys in June 1941, and the first to see action was the FCS HMS *Maplin*, a refrigerated freighter converted to an ocean boarding vessel. She was part of the escort for Convoy SL 81, which departed Freetown for Liverpool on 15 July 1941, carrying a Sea Hurricane Mk IA on its catapult, to be piloted by

Lieutenant R. Everett of 804 Squadron. When a Fw 200 was detected on 3 August 1941, Everett launched and attacked the Condor, severely damaging it, though he only narrowly avoided death himself on ditching the Hurricane nose first and which sank instantly. The Condor limped away but had to ditch off Brittany.

In fact, *Maplin*'s Hurricane had almost launched in the preceding convoy out to Freetown when F8+AB piloted by Hauptmann Fliegel began to orbit. The convoy's AA scored a somewhat fortuitous hit on Fliegel's aircraft, leading to it breaking up in mid-air, with the loss of that much decorated pilot and his crew.

The FCS *Springbank* took part in Convoy HG 73, which departed Gibraltar for Liverpool on 17 September, and was the first FCS to sail on a Gibraltar convoy. She was carrying a Fairey Fulmar two-seat fighter on her catapult, which

Poor quality but historically interesting images of NK+NM during a visit to southern Russia in the summer of 1941.

Both of the initial pair of FDF Condors, NK+NM in the background and 26+00 in the foreground, on 28 August 1941 when the aircraft carried Hitler and Mussolini to Uman, Ukraine, to visit troops taking part in operations on the Eastern Front.

was launched the day after departure at 1415 when a Fw 200 appeared. The Fulmar's performance was too low to decisively engage the Condor, but it was able to keep the bomber away from the convoy. Eventually the Fw 200 jettisoned its bombs astern of the convoy and made off, while the Fulmar flew to Gibraltar.

Over the next two days, a Condor appeared, but remained at a distance without carrying out an attack. *Springbank* was torpedoed by *U-201* on the 27th, and the next day no fewer than three Condors were spotted, one of which was noted "Making continuous homing signals and reporting the convoy's course and speed".[63]

Five FCSs were operated (four conversions plus HMS *Pegasus*, the RN's catapult trials and training ship) and 35 CAM ships. There were only a handful of combat launches of fighters from catapult ships but five Condors met their end as a result of catapult-launched fighters between August 1941 and July 1943, and they contributed much to the strengthening of convoy defences that turned the tide against the Condor.

A far more practical proposition though was the auxiliary aircraft carrier, later known as the escort carrier — a merchant ship hull and machinery converted with a flight deck to carry a small number of aircraft. The original proposal had been as an anti-submarine platform but the threats to convoys from aircraft, particularly the Fw 200, meant that the first examples of this type of vessel focused on fighters.

---

63   HMS *Wolverine*'s Report of Proceedings, Convoy HG 73, 2 October 1941, www.warsailors.com/convoys/hg73.html.

Three images of NK+NM at its home base of Rastenburg, East Prussia, close to Hitler's bunker.

The auxiliary carrier was first proposed at the same time as the CAM ship and was advanced almost as quickly. A captured German merchant ship was handed to the Blyth shipyard in January 1941 for conversion and commissioned in June, initially as HMS *Empire Audacity* (reflecting her previous Ministry of War Transport merchant ship name), soon shortened to HMS *Audacity*. Her aviation facilities were basic, with no hangar and aircraft simply to be stored on deck, and the complement was six Grumman Martlet (Wildcat) fighters of 802 Naval Air Squadron (NAS), to cover both air defence and anti-submarine duties.

Audacity's first convoy was OG 74 to Gibraltar in September 1941, and the carrier aimed to maintain a constant standing patrol of a pair of Martlets. The first encounter with a Condor came on 21 September when the rescue ship *Walmer Castle* and tug *Thames* were recovering survivors

from two ships torpedoed during the night. Leutnant Schaffraneck's Condor then bombed *Walmer Castle*, inflicting fatal damage. *Audacity* launched Black Flight, Sub-Lieutenants Patterson and Fletcher, who attacked as Schaffraneck was attempting to bomb the *Thames*. The Condor broke up and all the crew perished.

The convoy arrived at Gibraltar on 27 September, and *Audacity* returned with Convoy HG 74. Comprising half of Red Section was the then-Sub-Lieutenant E. Brown, later the test pilot known as 'Winkle' Brown. He records a lucky escape, when "On one sortie like this Red Section was boring its way up to a large four-engined aircraft which was already firing. We were still well out of range. Suddenly my little side windscreen shattered, and I felt a searing pain in my mouth and tasted blood." Brown wrote how he sustained further facial injuries on landing, catching the last wire and hitting his head on the gunsight.[64] This incident helped prompt 802 Squadron to develop different tactics, and in future a head-on attack would be favoured over the traditional quarter attack. If positioned carefully, a fighter could approach between the zones covered by the two forward-firing guns.

For future convoys, *Audacity* embarked an additional two Martlets, adding to the overcrowding but mitigating somewhat the high wastage rate experienced until then.

While the Martlets had undoubtedly been successful, there were problems with their armament that caused limitations. The Browning .50-inch guns had a tendency to jam, and in a report by the Admiralty's Anti-Submarine Warfare Division recorded that on 18 December: "During the forenoon two Focke-Wulfs were engaged by aircraft from AUDACITY, but unfortunately guns in both the fighters jammed. Although both Focke-Wulfs got away, one of them is believed to have been damaged."[65] The Fw 200s of Jope, now promoted to *Hauptmann*, and Oberleutnant Heindl, did indeed escape, to confirm that the convoy had a miniature aircraft carrier with it.

The following day proved more decisive, though showed that the Condor was still not to be trifled with. From the same report:

Soon after 1100 two Focke-Wulfs arrived in the vicinity of the Convoy. The first one was soon set on fire and shot down by two fighters (one of whom was Brown). In his report the Senior Pilot states — 'Hot return fire from the Focke-Wulf destroyed the pilot's hood, passing through where his neck should have been, fortunately he had got his neck near his boots just in time!'

HMS *Stork* confirmed: "The aircraft presently returned leaving a very dead Wulf." Another Fw 200 appeared at about 1600 and was shot down by Red Section "by means of a combined quarter and head on attack."

It was to be *Audacity*'s last hurrah. On the night of the 21st she was torpedoed by *U-751*, having quickly become a priority target. The concept of the auxiliary carrier had been thoroughly proved though, and things would only become more dangerous for the Fw 200s as time went on.

THE MEDITERRANEAN

In August 1941, a detachment of eight Fw 200s, six armed and two transports, and 10 He 111s from KG 40 transferred to Eleusis to work with X. Fliegerkorps in the Mediterranean at the behest of Fliegerführer Atlantik's Martin Harlinghausen. The so-called Suez-Kommando's purpose was to attack shipping in the Suez Canal and Red Sea. The first attack was made early the next month, when Leutnant Mayer bombed the merchantman *City of Auckland* by moonlight on 3 September in the Red Sea. Another attack was made, by Leutnant Dostlebe, on the 10 September, damaging the *Honduras* off Suez. This was the last shipping attack in this brief phase, the end possibly spurred by the loss of one of the aircraft, F8+GH, with pilot Oberleutnant Neumann and crew, 20 minutes after taking off on 5 September, and possibly by X. Fliegerkorps' evident dislike of the Fw 200. On 14 September a signal was intercepted from Eleusis to KG 40 at Bordeaux, noting, "Up to the present 3 Fw 200 A/C and most of the equipment have been sent to Bordeaux via Lechfeld."

In fact, it seems that Harlinghausen may have had a further plan. Harlinghausen's experience in the Spanish Civil War had given him great faith in the air-launched torpedo as a weapon against commerce. Again, six aircraft from 9./KG 40 transferred to the region, these having been modified for

64   Captain E. Brown, *Wings on my Sleeve,* Weidenfeld & Nicolson 2006, pp. 28–29.
65   Operations by HMS *Audacity* in defence of convoy HG76, 14–21 December 1941, Report by Anti-Submarine Warfare Division, 30 January 1942, ADM 199/1998.

Wounded troops from the first months' fighting against the Soviet Union in 1941 carried home for medical treatment in NK+NM, by personal order of Hitler.

torpedo carriage and were dispatched to Catania, Sicily, to train and prepare for torpedo operations. The A-0 F8+EU *Friesland* and D-2a F8+BU were employed as support aircraft to the detachment.

Considerable effort was expended in vain during 1941—42 to develop a capability for torpedo attack. Adapting the Fw 200 for torpedo attack was realistically the only possibility for a long-range capability with this weapon. There were several pressing reasons for attempting it. The experience of the Luftwaffe's Italian counterpart, the Regia Aeronautica, in the Mediterranean demonstrated the potency of a strong torpedo capability as part of maritime strike forces.

In late June 1941, torpedo training of KG 40 crews took place on the ranges at Grossenbrode. A British intelligence report discussing this noted, "It was found that the Italian torpedoes were superior to the German which often broke up on hitting the water and therefore it is intended to use the Italian 'Fiume' torpedoes of 750 kilos."

The German F5 series was an imperfect weapon, due to pre-war neglect of the aerial torpedo, with mediocre speed and range, poor reliability, and initially a maximum launch speed so low that it precluded the torpedo's use by modern monoplanes. Significant progress made in the first years of war with the improved F5a and F5b, but production was slow and the situation was so unsatisfactory that the Italian Fiume (Whitehead) F200/450 x 5.46 was adopted as an expediency, designated F5w. Over a thousand of the Fiume torpedo were acquired.

Interrogation of a captured Fw 200 crew in August 1941[66] revealed some details of operations with torpedoes:

The crews were taught to approach their target at about 260/280 kph, flying at 45 ft and rising to 150 ft, to release the torpedo. The aircraft flies directly at the ship, the angle of lead being allowed for by

66   Air Ministry, Directorate of Intelligence and related bodies: Intelligence Reports and Papers, AIR 40/154, Fw 200 Aircraft.

an adjustment on the rudder gyro of the torpedo.

After releasing the torpedoes, the pilot climbs and banks away towards the stern of the vessel, and on straightening out switchbacks.

Under ideal conditions, crews try to release the torpedoes from a range of 800/1,000 yards. The minimum range is 500 yards, and the maximum for accurate aiming, 1,800 yards.

Flieger-Oberstingenieur J. Hillerman, described as 'Chief of German Aircraft Experiment', wrote that the procedure for sighting a torpedo attack was "very inexact and rough".[67]

The interrogation report noted that "A box-like air fin was fitted to flight the torpedo through the air. This was said to have three parallel horizontal vanes".[68] This was the K3 tail which was fixed and merely a passive stabilisation device. "The release of the K3 tail had not always gone well," according to Hillerman, and if the torpedo did not enter the water straight, it could fail to break away. The F5b did, however, have an 'anti roll ring' with horizontal fins that operated in the first second after the torpedo entered the water to counteract any rolling motion — though this introduced numerous problems of its own and was withdrawn before 1942.

The captured crew indicated that torpedo-carrying He 111s could adjust lead and depth of the torpedo in flight, but on Fw 200s these parameters had to be set on the ground. They reported that the torpedoes were set for a depth of two metres and lead calculated for a ship steaming at 8 knots, quoted as 2°. The torpedoes were set to spread somewhat, with one set to steer to starboard and the other to port. The crew confirmed

that carrying torpedoes necessitated removing a section of landing flap which would otherwise have impinged on the torpedo. Carrying torpedoes reduced the speed of the Fw 200 by about 10km/h.

Developments had evidently been made since April, as the report noted: "The depth at which they are to run and the rudder angle can be set on to the torpedoes from inside the aircraft whilst in flight." It added: "For range estimation an instrument known as the 'Entfernungsschätzungsgerät' may be used. This simply consists of two adjustable vertical pointers which can be set according to the size of the ship. The aiming will probably be done by the pilot, using a Revi sight."

Mediterranean torpedo operations, however, never took place. The unsuitability of the Condor for torpedo attack was one likely reason, another being X. Fliegerkorps' reluctance to support the type. An intercepted signal between X. Fliegerkorps HQ and Fliegerführer Afrika on 7 December stated: "It is impractical to employ a Condor *Staffel* here as ... the technical maintenance of such special aircraft could not be guaranteed, as the strain on our workshops — responsible for operational and many transport units — would be too great." The signal added that "Condors are not suitable for carrying barrels and are no advantage when it comes to transporting troops". It transpired that X. Fliegerkorps had "on its own initiative transferred the *Staffel* of Oberleutnant Ickenroth to support Italuft's[69] efforts." The Condors were moved to Lecce in Italy where they began carrying out transport flights. Despite X. Fliegerkorps' low opinion, the Condors seem to have been of considerable use. An intelligence report from May the following year included that prisoners of war "Stated that Rommel's advance in January 1942 was probably made possible by

67 J. Hillerman, Flieger-Oberstingenieur, A Comprehensive Study of the Development of the German Aircraft Torpedo from 1942 to 1945, ADM 213/214, pp. 5–6.
68 Fw 200 Aircraft, Air Ministry, Directorate of Intelligence and related bodies: Intelligence Reports and Papers, AIR 40/154.
69 *General der Deutschen Luftwaffe beim Oberkommando der Kgl. Ital. Luftwaffe* – the HQ of Luftwaffe flying units in Italy.

the ferry service of petrol supplies from Catania to Aral Philaenovum run by Fw 200s, five of which made three trips a day, each carrying each time 15,000 litres of petrol".[70]

The Condors returned to Bordeaux in late December 1941 or early January 1942, though there were occasional forays back into the Mediterranean, usually for transport. For example, on 4 May 1942, D-2 W. Nr. 0019 *Rheinland*, then operating as a training and transport aircraft with 8./KG 40, was dispatched to Lecce to support IV./KG 40 with transport flights to North Africa. The flights continued until 0019 was lost in a thunderstorm near Zante on 2 June.

A small number of sorties carrying torpedoes on the Fw 200's usual hunting ground were made in the last days of 1941. An armed reconnaissance on 13 December from Rennes found no targets, and an attack was made on an unescorted steamer on 30 December west of Portugal, without result.

Some accounts suggest a further operation may have taken place in early January 1942. Interestingly, a report from an unidentified Naval Armed Guard aboard a merchantman suggests a torpedo-armed Condor attacked a convoy:

At 0440, four 4-motored Junker 90 or Focke-Wulf airplanes came upon the convoy off the port quarter and flying within 150 feet of the water. Evidently the two escort vessels guarding the port flank did not see the planes, because not until after the two after .50 caliber machine guns and two .30 caliber opened up did the escort vessels commence firing. The concentrated firing from our guns diverted the planes, and kept them from flying directly over the convoy, causing them to drop their torpedoes which passed harmlessly by our stern.[71]

Fw 200 crews continued to be trained in torpedo attack as late as January 1943, but by August 1942, it was apparent that the torpedo was not a routine weapon for the Fw 200. "The Fw 200 has actually been used for long-range torpedo work, but it is not really suitable for this and is unlikely to be used extensively,"[72] a report on the 7th stated. Indeed, J. Hillerman's 1945 report did not even include the Fw 200 in its table of 'German torpedo-carrying aircraft'.

---

70   M. E. Letter 214 C.S.D.I.C. 152 29/5/42 — FW200, Intelligence Reports and Papers, AIR 40/154.
71   Armed Guard Voyage Narratives, General Instructions for Commanding Officers of Naval Armed Guards on Merchant Ships 1944, Navy Department, OPNAV 23L-2, pp. 150—51.
72   Fw 200 Aircraft, Air Ministry, Directorate of Intelligence and related bodies: Intelligence Reports and Papers, AIR 40/154.

4

1942

ATLANTIC DECLINE

The increased difficulties on the Atlantic sea lanes and the opening of a new front at sea with convoys to Russia led to a change early in 1942. I./KG 40, now under the command of Hauptmann E. Daser, moved from France to Vaernes, near Trondheim in Norway. Dönitz had actually requested this the previous year and been denied, but now it was forced by circumstances. The success of HMS *Audacity* during HG 76 led to five more merchantmen being released for conversion to escort carriers.

It was fortunate for the Allies that *Audacity* had changed the game when she had, as despite the continuing organisational difficulties, cooperation between Condors and U-boats had been working well and achieving significant results. A system had been established whereby if a U-boat tracking a convoy had to submerge, it would be informed certain times at which a Fw 200 would reconnoitre so the submarine would be primed to receive the intelligence and re-establish contact with the convoy. B.d.U. would listen into aircraft radio transmissions and then pass on the frequency to U-boats for homing signals. Direct homing meant that the problems of insufficiently accurate navigation were unimportant as the submarines were following a signal rather than relying on a reported position. This was further assisted by the Schwan airdropped VHF beacon.

The reorganisation changed that dramatically. III. Gruppe was converting to the Condor from the He 111, and remained in France, but was focused on training and continued to suffer from poor serviceability. Few Condors troubled the Atlantic sea lanes in the early part of 1942. Perhaps unsurprisingly, one of the more successful pilots during this period was Hauptmann Jope, who sunk two vessels in March (albeit one of them a neutral Portuguese trawler).

In truth, the Condor was now soldiering on beyond the point at which anyone had expected to need it to. The Heinkel He 177 had been touted as the answer to maritime strike when it first flew in 1939, but in early 1942 the scale of the programme's difficulties was all too apparent. Junkers was developing the Ju 290, an evolution of the Ju 90 transport, but the prototype was still months from completion.

Focke-Wulf would therefore continue to improve the Fw 200, with the C-4 variant. This started life as the *Umbau* 5 factory modification of the C-3 (C-3/U5) with internal fuel increased to 8,600 litres courtesy of a self-sealing tank in the forward part of the fuselage tub, further airframe strengthening and some improvements to the

Two images of F8+GH, W. Nr. 0074, one of the six Condors transferred to the Mediterranean for operations against the Suez Canal.

An unidentified C-4 with a white band around the outer wing, which could indicate service in the Mediterranean.

armament. Although it was not yet standard equipment, numerous C-4s would be fitted with the Lofte 7D bombsight (Carl Zeiss Lotfernrohr 7). The bombsight could not physically fit into the fuselage tub beneath the bomb aimer's position, so it was enclosed within a deep fairing on the underside of the tub. It was not possible to accommodate both the sight and a 20mm cannon, so aircraft equipped with the Lofte were armed with a smaller-calibre gun.

Despite the disruption to Fw 200 operations in early 1942, the procedures worked out during the Gibraltar convoys were put into effect for the Russian convoys and began to take effect. With more heavily armed merchantmen and plentiful escorts, and the typically poor visibility and rough seas on the northern convoys, it was less suitable for higher altitude attacks using the Lofte 7D, but the Condor once again came into its own as a reconnaissance platform for U-boats.

Convoys bringing supplies to the USSR from the US and UK began in August 1941, two months after Germany's invasion. It was essential to the western Allies that the USSR was not defeated, releasing men, raw materials and industrial might that could be turned against them. The PQ series of convoys to northern Russian ports started in September. They had to sail around the northern coast of occupied Norway, in range of Luftwaffe bombers as well as under threat from U-boats and surface warships. The Condor's range was less of an advantage than it had been over the Atlantic, but its endurance was of vital importance, for locating and shadowing convoys, and homing other forces onto them.

After I./KG 40 moved to Vaernes in March, it was soon in action. The westbound QP 10 was the first to receive the Condors' attentions, leaving Kola Inlet on 10 April. A Fw 200 (later relieved by a Bv 238) tracked the convoy, which was attacked by Ju 88s on the 11th and 13th, and a U-boat on the 13th. Later that month, a I./KG 40 Condor

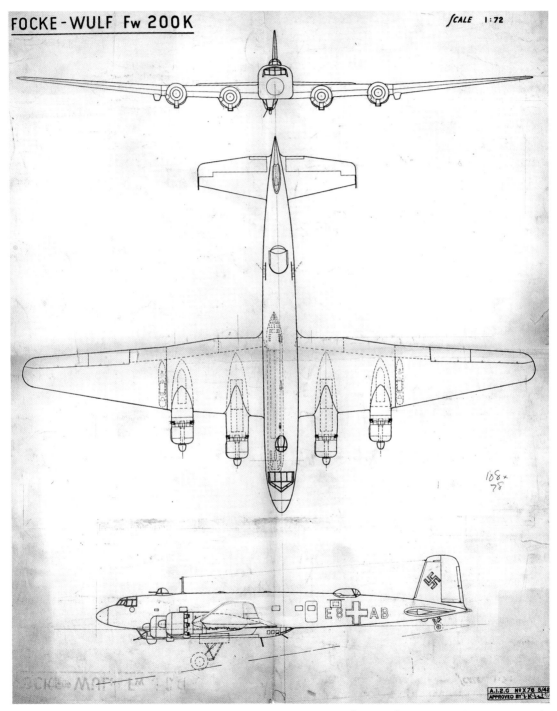

FOCKE-WULF Fw 200K

SCALE 1:72

British Admiralty intelligence produced this fairly accurate drawing of a 'Fw 200K' – a spurious designation merely indicating a military version – in May 1942 for identification purposes.

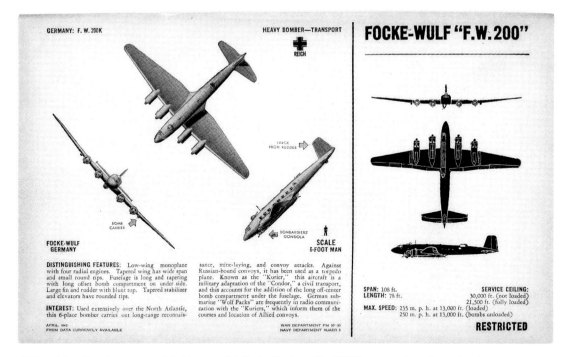

US Navy identification poster of the Fw 200, chiefly for the benefit of US Navy escorts and merchant gun crews in the Atlantic. The left-hand three-view fails to include the kinked trailing edge that had been a feature of the Condor since 1938.

successfully located the Russia-bound PQ 15 south-west of Bear Island on 30 April, and attacks by torpedo bombers and a U-boat led to the sinking of three merchantmen. Then, on 13 May, another of the *Gruppe*'s aircraft spotted a group of ships attempting to shepherd the damaged cruiser HMS *Trinidad* back from Russia, and the subsequent air attack led to *Trinidad*'s abandonment and scuttling.

It was PQ 16, which departed Reykjavik on 21 May, that really proved the adeptness of I./KG 40's Fw 200s. Eight of the *Gruppe*'s Condors, helped by the almost constant daylight, worked in relays to provide ongoing homing signals for U-boats and aircraft. Seven merchant ships were sunk. The following PQ 17 at the end of June was disastrous for the Allies, with fewer than a third of the ships making it through, though even dispersed and disrupted, the convoy proved fatal

for Oberleutnant Gramkow whose Fw 200 C-4 was shot down by SS *Bellingham* in the Barents Sea. The next PQ convoy would not sail until September when an escort carrier could be provided – HMS *Avenger*, with a complement of Hawker Sea Hurricanes. A Fw 200 spotted the carrier while on reconnaissance of Iceland and tried to bomb the carrier in Seyðisfjörður, while the convoy was still assembling, but missed. This was the only contact I./KG 40's Condors had with this convoy, and then the Russian convoys were suspended until December.

Regular reconnaissance of Iceland was one of I./KG 40's tasks during this period and it, too, was not without its risks. On 24 October 1942, Oberleutnant Godde and his crew were flying over the western part of the island when they were intercepted by USAAF Bell P-39s from the 33rd Fighter Squadron, which had been based on the island for

Recognition photograph of a model built for identification purposes. This model or one very like it was used in a film made the Directorate of Army Kinematography by Analysis Films showing how to recognise the Condor. A comparison photo of a Fw 200 indicates that while the model is not entirely accurate, it captures the main visual points.

KG 40 crews walking back from their aircraft at Bordeaux-Mérignac – probably a posed photograph given the neatly lined up Condors.

its defence from late 1941. The P-39s were flown by Second Lieutenants Ingleside and Morrison, who made repeated attacks on Godde's C-4/U3 until it was seen to be on fire and losing height before crashing on Arnarvatnsheiði, a plateau of high ground where remains of the Condor can still be seen. The wreckage was studied by the Allies at the time, and though it was "badly destroyed", it furnished some useful information, such as that the aircraft carried a radio altimeter, as well as a complete MG 151/20 20mm cannon for study. Evidence that one gun of this type had been mounted in a fixed, forward-firing position was of particular interest.[73]

The day after this incident, a number of Condors were again transferred from III./KG 40 to Lecce to transport fuel for the Afrika Korps, as they had a year earlier.

In December, the opportunity was taken to employ Condors for an attack on shipping assembling in Casablanca harbour. Twelve aircraft were to take part, but one failed to leave Bordeaux and three more had to turn back because of technical difficulties. One more failed to find the target, and of the seven that bombed Casablanca, none caused much damage. Two Condors then had to divert to Spain for a forced landing, one of them crashing in the sea with the loss of the crew. It was not an auspicious episode.

VIP OPERATIONS LATER IN THE WAR

From June 1941, operations by the Fliegerstaffel des Führers were split. A forward element was based near Hitler's bunker at Rastenburg, East Prussia, while the rear element remained at Tempelhof. Naturally, the forward element typically included

---

73   Crashed Enemy Aircraft, Report Serial No. 159 dated 14 November 1942, Focke-Wulf 200, Intelligence Reports and Papers, AIR 40/154.

KG 40 personnel demonstrating different levels of ability when manhandling 50kg bombs while loading up an aircraft before a flight.

26+00 and NK+NM. With the latter being officially designated the escort to 26+00, it is unsurprising that the two aircraft were seen together on a number of occasions. One such instance was a joint visit to troops on the Eastern Front by Hitler and Mussolini following the Axis victory in the Battle of Uman in August 1941. Mussolini travelled in 26+00 from Rastenburg with Hitler and was photographed seated in the cockpit alongside Hans Baur.

NK+NM's time in this role was relatively brief, and it was written off before the end of the year. On 23 December, NK+NM was dispatched back to the Eastern Front to deliver Christmas gifts from Hitler to troops taking part in the invasion, with Flugkapitän Ludwig Gaim at the controls. While landing on a snowy runway at Orelin, Gaim lost control and the Fw 200 crashed. NK+NM was written off, its back broken. Gaim survived, albeit with fairly severe injuries.

The other Fw 200 of this pair, 26+00, continued to operate with the FdF for most of the war. Nevertheless, Hitler's personal Fw 200, 26+00, began to fall out of favour when armed Fw 200 VIP transports started to become available in January 1942, although it continued to bear his unique call sign for the remainder of its existence. The *Führer* preferred to use the armed C-4/U9 when that machine appeared in winter 1942, but 26+00 remained in use, carrying figures such as Reichsmarschall Hermann Göring, in the winter of 1942–43, and Großadmiral Dönitz during a visit to Rome to meet Italian admirals in May 1943.

As the war progressed, and Allied long-range fighters became more of a threat, the FdF took on several armed Fw 200s. These were adaptations of the military C variant, but with unique combinations of armament and equipment. They included the sole C-3/U9 W. Nr. 0099, delivered in January 1942, and the pair of C-4/U1 aircraft, W.

Lifting an SC250 bomb into the outer nacelle recess bomb rack.

Carrying MG 15 machine guns aboard F8+CH at Bordeaux-Mérignac. Note the small slots beneath the door for the ladder to hook into.

Unidentified crewmember departing from a KG 40 Fw 200. A C-1/2 can be seen in the background.

A member of KG 40 aircrew is greeted by officers after a successful flight, presumably having reached an important milestone, as others disembark.

Contemporary colour image of a Fw 200 C-3.

Nr. 0137 and 0176, delivered in January and May 1943 respectively. The C3/U9 was coded KE+IX and retained this identity despite KE codes generally being associated with temporary delivery identities. The first of the C-4/U1s was coded CE+IB, and the second became GC+AE. Both were used by Hitler at various times — the first was earmarked for the Führer while the other was intended for Reichsmarschall Göring but ended up being assigned to Reichsführer-SS Heinrich Himmler. A third machine of this sub-type, W. Nr. 0240, joined the FdF in September 1944 as another aircraft for Hitler, with the code TK+CV.

The C-3/U9 differed from the contemporary military C-3s in its defensive armament and the deletion of all bomb carriage, although it retained the long tub of the military version. It could carry 11 passengers. While military variants had mostly switched to the larger, powered DL15 turret in the A-stand position, the U9 reverted to the D30 manually operated turret.

In addition, a pair of C-4/U2s were adopted by the FdF, W. Nrs. 0138 and 0181 as *Begleitflugzeuge* ('escort aircraft') — the latter became the personal aircraft of Großadmiral Dönitz named *Albatros III* until the end of the war. These had a unique short fuselage tub. Two C-6/U2s and a C-4/U1 were received in 1944. A total of 10 Fw 200s would ultimately serve with the FdF.

Luftwaffe *Major* in white tunic talks to civilians at Bordeaux-Mérignac in front of a Fw 200. He has the Luftwaffe Reconnaissance Flying Clasp issued for completing a certain number of operational flights above the left breast pocket, and the German Cross on his right breast pocket. The officer next to him is wearing the War Merit Cross medal ribbon.

KG 40 Fw 200 C-3 at Vaernes, Norway.

Two studies of a Fw 200 C-3 in flight. The first clearly shows the partially fabric-covered wing surface in the outer panels.

A Fw 200 C-3 of I./KG 40 coming in to land at Vaernes.

Fw 200 C-3 showing evidence of winter camouflage in the show at an airfield in Norway.

USAAF personnel with the remains of the tail cone and elevator of the Fw 200 C-4 shot down over Iceland on 24 October 1942.

Two photographs of Hitler's Fw 200 26+00 in
temporary winter camouflage.

5

1943—45

In late November 1942, Soviet forces successfully encircled the German 6th Army at Stalingrad. Measures were immediately taken to implement an air-supply operation of the kind that had been seen in Norway back in 1940. The transfer of 20 Condors from KG 40 where they had been having a material effect on the vital Battle of the Atlantic was an indication of the priority afforded to the operation.

According to Generalmajor Friedrich-Wilhelm Morzik, the head of Luftwaffe Transport:

At the order of the Quartermaster General, all Ju 52s, Ju 90s, Fw 200s, He 111s, and even Ju 86s (though totally unsuited for transport purposes) were requisitioned from all units, staffs, ministries and — as usual — the Office of the Chief of Training. The fact that 600 aircraft, together with some of the best flight instructors, were commandeered from the latter source alone, is ample illustration of the ruthlessness with which every available aircraft was mobilised for the air-supply mission at Stalingrad.[74]

A new special duties unit, KG z.b.V. 200, was set up under 11./KG 40's Major Willers, according to Morzik "equipped with Fw 200s (Condors), Ju 90s and two Ju 290s. The unit appeared reassuringly in the daily morning reports sent out to the Luftwaffe High Command and to other top-level headquarters, but its appearance over Stalingrad as part of a supply mission was rare indeed."

The conditions restricted the Condors to a low sortie rate. At first, supplies were flown in, then dropped by parachute. After a move to Salsk after Tatsinskaya was evacuated on 24 December, "The approach route was now nearly 250 miles long, dangerously close to the maximum range of the transport aircraft."[75] Generalmajor Morzik went on to what faced the Fw 200 crews in Russia:

There was no time for the crews to become accustomed to conditions gradually; they had to be employed fully, right from the start. The harshness of the winter cold; the long and dangerous approach and return flights over enemy territory in the face of heavy fighter and antiaircraft artillery defences; the constant enemy bombardment of take-off and landing fields; the

---

74   Generalmajor a. D. Fritz Morzik, German Air Force Airlift Operations, USAF Historical Studies No. 167, USAF Historical Division 1961, p. 186.
75   Ibid. p. 191.

Condors engaged for the Stalingrad and Kuban airlifts November 1942–February 1943 had to endure cold temperatures and basic conditions that hampered serviceability, though things improved considerably for the Kuban operations.

Fw 200 C-4 with defensive armament removed, probably in use in the transport role in 1942–43.

unloading and loading in the encircled area, while constantly harassed by artillery fire and grenades; and the ever-present danger of icing and other technical failures in the bitter cold all combined to make the situation faced by the transport personnel a very difficult one.[76]

The last pockets of resistance collapsed on 2 February. Fw 200 losses during the Stalingrad operations totalled nine lost or damaged. The numbers of Fw 200s involved were tiny compared with the overall air fleet – losses of Ju 52s totalled 266 and of He 111s, 165 – but the number of airframes lost or damaged represented almost half the number taking part. Eighteen crewmembers were posted missing, with five wounded. A C-3/U2, the former F8+GW of I./KG 40, was captured by

Russian forces on 31 January, repaired and tested.

The German collapse at Stalingrad prompted a new Soviet drive westward, which in turn cut off German forces in Kuban in the Caucasus. The Kuban bridgehead necessitated another airlift, by forces in no position to repeat the immense efforts at Stalingrad. KG z.b.V. 200, by now "exclusively equipped with Fw 200s",[77] was enlisted to help carry out individual air-supply missions, as well as bombing missions against Soviet rail targets. Großadmiral Dönitz attempted to order the Condors back to maritime duties, but this was countermanded by Generalfeldmarschall von Richthofen and KG z.b.V. 200 was assigned temporarily to Luftflotte 4, "To deliver supplies of ammunition, gasoline, and food to the Kuban Bridgehead. On the return flights all available space was to be utilised for the transport of

76   Ibid. p. 188.
77   Ibid. p. 204.

Scene from the Stalingrad airlift in 1943 at a wrecked aerodrome with a Ju 88 from KG 51 'Edelweiss' and a KG z.b.V. 200 Condor still wearing KG 40 markings behind.

wounded personnel".Generalmajor Morzik noted that "The group itself, whose personnel were the same as during the Stalingrad operation, was fully qualified. The crews had had a great deal of experience and their flying ability was wholly adequate to the demands of the operation".[78]

Circumstances allowed for better results this time too. A detachment of specialist mechanics who had been requested weeks before to support the Stalingrad airlift arrived from IV./KG 40 on 4 February, along with spares and special equipment, immediately improving the serviceability of the aircraft.

The fact that Fw 200s made up the entirety of the small operation, rather than a tiny part as in the Stalingrad efforts, meant that activities could be tailored much better to suit the aircraft. Morzik reports that "Whenever instrument landings were

to be made, the group commander saw to it that the direction finding station was manned by an experienced group radioman, for only someone from the group itself could be thoroughly acquainted with the abilities of the individual crews and the flight characteristics of the Fw 200". Due to the experience of the crews with weather reconnaissance, the transport flights also provided valuable weather information for other units.

Flights to the bridgehead began on 4 February, when seven Fw 200s transported supplies from Zaporozhe to Slavyanskaya. There were also two missions between Krasnodar and Kerch, returning with wounded personnel and even "a load of copper". The pattern continued for the next nine days, always during daylight, with other missions to transport ammunition from Zaporozhe to

78   Ibid. p. 205.

FdF Condor C-3/U9 KE+IX, the first armed FDF Condor, at HQ 'Wehrwolf' in Winniza, Ukraine, during the winter of 1943.

Timashevskaya and transfer wounded personnel to Crimea. Additionally, on each night the Fw 200s bombed Soviet rail targets. On 9 February, the order was received to transfer to Berlin-Staaken, but transport missions continued until the 13th.

Ultimately, while there were still good reasons for continuing supply missions to prepare the forces in the bridgehead for the inevitable battle with the encroaching Soviet forces, the vulnerability of the valuable Fw 200s to enemy action or accidental loss – a minor technical failure might mean the loss of the aircraft due to the difficulty in receiving spares – meant that the risks were considerable. Morzik noted that "The Fw 200 might have been better employed in bombardment or armed reconnaissance missions far into enemy territory, for it was well equipped for utilisation as a long-range reconnaissance aircraft". Indeed, the *Gruppe* was retained in the east until

22 February engaged in offensive missions, but the aircraft and crews were now in great demand for a revitalised Atlantic campaign.

LAST OCEAN OPERATIONS

Maritime operations were much reduced in early 1943, with so many Condors diverted elsewhere for transport operations. Indeed, there were no combat losses on patrol operations until March. Accidents and even friendly fire would be a greater threat until anti-convoy missions over the Bay of Biscay began to pick up in April and May. KG 40 hoped that new equipment, tactics and training would help bring new success, and repeated massed attacks were planned for the first time since 1941.

The Allied commitment to countering the Condor threat with fighters was unabated. While the Luftwaffe attempted to provide support with

KE+IX taxiing on the Eastern Front later in 1943. This aircraft like all the armed FdF Condors wore wraparound black engine nacelles, presumably as a recognition feature.

Ju 88 fighters, long-range Beaufighters were now ranging over Biscay and new US-supplied escort carriers equipped with Seafires and Wildcats to replace the old, under-armed Sea Hurricanes were increasingly available. One such was HMS *Battler* which left Belfast on 4 June 1943 escorting Gibraltar convoy OS 49. Twenty days into the voyage, the carrier's radar detected a 'bogey' which proved to be Fw 200s. Sub-Lieutenant Gordon Penney, a Seafire pilot with 808 Squadron, recalled:

Peter Constable and I were on stand-by when the convoy was attacked by Focke-Wulf four-engined bombers just a few minutes before sunset. We were airborne in no time and Constable asked me to

lead as he had lost sight of the enemy aircraft. I could only see one and climbed to a position above and astern of it … About halfway through my attack, tracer fire from the rear gunner ceased and, breaking away, I saw that the port wing of the aircraft was ablaze.[79]

Oberfeldwebel Abel's C-4 was never seen again, and nor were any of the crew.

On 11 July 1943 a decrypted signal from a U-boat at 2051 noted three merchant ships with three escorts 350 miles west of Porto — the small Convoy 'Faith'. Later that evening, the same U-boat reported that the Luftwaffe "Reports hit on 20,000 tonner and another steamship in naval

_____

79   Penney, Lieutenant (A) AG; Flying Navy 1940–1944 (unpublished manuscript), p. 27.

Fw 200, possibly an FdF aircraft, on the Eastern Front.

grid square CF 3651. Both steamships are on fire".[80] This was part of a renewed effort by III./KG 40 to make direct attacks on shipping, the two vessels mentioned being the troopships *California* (16,792 GRT) and Duchess of York (20,021 GRT). A second attack accounted for the third vessel, the stores ship MV *Port Fairy* (8,072 GRT).

The attack had been enabled by the Lofte 7D bombsight, which III./KG 40's Fw 200 C-4s had begun to have fitted from the summer of 1942. This conferred the ability to bomb with great accuracy from medium altitude. The results, relatively speaking, were spectacular. So rare was it for Condors to attack directly by this time that the convoy's commodore had elected to keep the ships in line abreast formation, believing that the Fw 200s would merely direct submarines into the

attack. The results caused an immediate response by the Allies, who were forced to extend the line of north–south convoys considerably to the westward. A new variant of the Condor, the C-5, introduced in 1943 was the first to fit the 7D as standard equipment.

The Lofte 7D also enabled attack with a revolutionary new weapon – the Henschel Hs 293 radio-controlled glider bomb, generally referred to simply as '*Kehl*' in German sources after its guidance system. (A dedicated version, the C-8, would be produced that could operate this weapon without modification, but other variants could be adapted to use it.) This promised to bring a new lease of life to the Condor, as it could deliver a devastating attack against shipping from outside the range of AA fire, and there was very little that

80   Teleprinted translations of decrypted Second World War German U-boat (or U-boat command) radio messages, German and Italian U-boat activity and organisation during the Second World War, HW 18/266.

Fw 200 C-3 under guard. The D-stand is fitted with a 20mm cannon.

could be done to stop it. The bomb was guided from the aircraft and a red flare enabled the 'pilot' to make course corrections in flight. A report from a merchantman in November 1943 reported that a "Fw (?200) released glider when flying across the wind and convoy course both ahead and appeared at about 3000 feet distant 2½ miles", further noting that "Guiding red light most conspicuous".[81]

The FuG 200 Hohentweil and FuG 217 Rostock search radar was also fitted to the Fw 200 from 1942. These allowed surface ships to be located at night or in poor weather, and aircraft fitted with them could be identified by the antennae sprouting from the aircraft's nose. The C-5, C-6 and C-8 were typically fitted with radar.

The *Kehl* glider bomb could only be used in

"slight, high or medium high cloud", according to the Fliegerführer Atlantik staff,[82] which recommended that aircraft "be fitted also with torpedoes". As a result, work continued to develop the Fw 200 for torpedo use into 1943. This is apparent from the loss of Fw 200 C-3/U1 DE + OG, which received 25% damage during torpedo trials at Lecce on 7 January 1943, and crashed the following day with the loss of Oberleutnant G. Ulrich on its return to Toulouse. Information on losses reveals further periods of Condors operating in the Mediterranean on occasion in 1942 and 1943, but this was largely for training and transport.

Towards the end of 1943, the U-boat force was finding it increasingly difficult to locate convoys, throwing an even greater emphasis onto aerial

81   Signal S.O. 40th E.G. WX 69258 22/11/43, Intelligence reports and papers, AIR 40/154.
82   Headquarters Staff Fliegerführer Atlantik, Principles Governing the Conduct of Operations by Fliegerführer Atlantik and an Appreciation of the Types of Aircraft Available, 3 December 1943.

Armed FdF Fw 200 C with a crowd and staff cars gathered around it during a VIP visit.

reconnaissance. "The difficulty of finding convoys must be removed through far-reaching air reconnaissance on our part with location gear," stated a message sent out to all U-boats in November, 'location gear' referring to the radar then fitted to most maritime aircraft. Although it noted that the Ju 290 was becoming available, with the ability to penetrate 1,400 miles from its operating bases, "in addition to these, we have Bv 222s and Fw 200s". The message referred to "The fiasco of the past month in the Atlantic" which "has its roots in the failure to find the convoys". Rather forlornly, the signal added, "Though your fight may seem at present to offer little success for you, it is your

homeland you are protecting."[83]

In December, Fliegerführer Atlantik recorded

the Fw 200 is at present available in three different models:

a) Normal Fw 200 with a radius of 1500 km.
b) Fw 200 fitted with auxiliary fuselage tanks (known as long range Condor) with a radius of action of 1750 km.
c) Fw 200 fitted with auxiliary fuselage tanks and two exterior tanks (known as maximum range Condor) with a radius of action of 2200 km).

83   Teleprinted translations of decrypted Second World War German U-boat (or U-boat command) radio messages, German and Italian U-boat activity and organisation during the Second World War, HW 18/266.

Fw 200 C-6 F8+AD of Stab. III./KG 40, demonstrating the FuG 200 Hohentweil radar aerials on the nose, during fuelling in 1943–44.

Only the long range and the maximum range Condor are suitable for the present operational commitments of Fliegerführer Atlantik. Use of the maximum range Condor is limited due to the difficulties involved in taking off at night because of overloading and its use can only be recommended for major operations. In view of its inadequate armament and its lack of speed the Fw 200 cannot be used in areas covered by enemy T[win] E[ngine] fighters. Recent encounters between Fw 200s and enemy TE fighters when cloud cover has been insufficient have nearly always led to the destruction of the Fw 200.

Further development of the Fw 200 is not recommended since:

a) it has been exploited to the limit of its potentialities,

b) it is being replaced by the He 177.

The fears of Fliegerführer Atlantik relating to Allied heavy fighters were well founded. The adoption of the Bristol Beaufighter and latterly the de Havilland Mosquito in Coastal Command service meant that the Condor simply could not operate safely where these long-range, powerfully armed aircraft were able to reach. This included well into the Bay of Biscay. A flight of five Condors was proceeding to attack Convoy OS 67 on 12 February 1944 when they encountered a patrol of 157 Squadron Mosquitoes, reported by the British squadron thus:

When at 45N 0722W 5 Fw 200 flying 280° at 0 ft in V formation were seen. Mosquitoes attacked from starboard. F/Lt Doleman and F/O Hannawin reformed and attacked again and E/A was seen to hit sea at 16.45

Fw 200 C-8, showing the FuG 200 Hohentweil radar aerials and the lengthened fuselage tub, which on this version extended further forward.

and blow up. F/O Hannawin saw strikes on a second A/C in second attack.

It was little safer in northern latitudes where after struggling to spare escort carriers for Russian convoys between September 1942 and September 1943 due to other commitments, the Royal Navy was once again providing strong fighter cover: two FAA pilots from 842 Squadron, HMS *Fencer*, both on their first operational flights, downed a Condor within minutes of launching, despite the kind of thick weather that was usually the metier of the Condor. *Fencer* was part of the escort for Operation *Tungsten*, a raid on the battleship *Tirpitz* holed up in a Norwegian fjord, so the destruction of the Condor on 22 March undoubtedly helped maintain the operation's secrecy.

Sub-Lieutenants Fleischmann-Allen and Charles were both praised by Admiral Max Horton, Commander-in-Chief of the Home Fleet, for their quickness of response to the Fighter Direction Officer, skill, and determination in pressing home their attacks in "the bad visibility and low cloud conditions prevailing" in which "the enemy might well have got away. As it was, the enemy, a Fw 200, was shot down seven minutes after take-off..[84]

The game was clearly up for the Condor, which by this time was well into its replacement by the Ju 290 and He 177. In addition to the Condor's difficulty surviving engagements with enemy aircraft, that month III./KG 40 reported only eight of its 35 aircraft were serviceable. While operational losses were falling, largely due to the reducing number of operations, losses from accidents were mounting. III./KG 40 gamely attempted to maintain weather reconnaissance flights for a while, before the

---

84   HMS *Fencer*, Enemy aircraft destroyer March 1944, two awards to FAA personnel, ADM 1/29501.

Fw 200 C-8 TO+XL, W. Nr. O256.

Gruppe was condensed into a transport squadron, the Transportstaffel 'Condor,' and Fw 200 combat operations ceased.

This unit continued operating in the transport role with a dwindling collection of aircraft throughout 1944 and into 1945 as German forces retreated on all fronts. A note from OKL to the Luftwaffe Organisationsstab on 3 March 1945 signalled the end for the Condor in Luftwaffe service. The recommendation had evidently come from the Transportstaffel 'Condor' itself:

It is proposed that the FW 200 that are still available be gradually supplied to DLH and handed over to them. The crews with good sea experience that become free could be made available for the western reconnaissance units ... The proposal made by the Condor transport aviation

squadron is considered to be very useful, since the necessary maintenance work can no longer be guaranteed under the current conditions, and the aircraft are therefore worthless to the Luftwaffe.

Gradually, as if by a form of natural reversion, surviving Condors were making their way to DLH. In March 1944, the Luftwaffe general staff ordered KG 40 to transfer four Fw 200s to DLH for the regular service to Spain. All were destroyed in Allied air raids, one in May during its conversion into an airliner at the DLH maintenance centre, and the rest in August. The old S5, W. Nr. 2995 *Friesland* went back to DLH in July 1944, and operated with the airline until 29 November when it crashed in the Baltic with the loss of all crew and passengers, possibly shot down by accident by a German patrol boat. This was the last of the original A-0

Captured example of the Henschel Hs 293 *'Kehl'* glider bomb, which could be launched by Fw 200s.

series. That month, three Fw 200 Cs were allocated to DLH to replace previous losses and were adapted to airliner standard. These were given the serials D-ASHG, D-ASHH and D-ASVV. The first two were given the names of previously lost DLH Condors, *Grenzmark* and *Hessen* respectively.[85]

On the death of Hitler, the FdF was renamed Kuriergruppe OKW. After the unit had to withdraw from Rastenburg, elements operated from Pocking, Bavaria, along with one of the last DLH Condors, D-AABR *Rheinland*, which was eventually captured by US forces.

---

85   Some sources state that the V10, W. Nr. 0001 was initially registered D-ASHH before its transfer to the Luftwaffe in 1939. This is likely confused by the fact that the aircraft was also allocated the name *Hessen* for service with DLH, but it is unlikely that the civil registration would have been the same too, not least as the new *Grenzmark* was issued the preceding serial.

Late-war line-up of Fw 200 C-8s. The second aircraft, C-8/U10 F8+FR, is wearing the *Mäandertarnung*, or 'meandering camouflage' with an upper splinter pattern of RLM 70 and RLM 71 over RLM 76, with a *Wellenmuster* or 'wave pattern' in RLM 76 applied over the top.

Factory-fresh Fw 200 C-4 W. Nr. 0097 before delivery to 3./KG 40.

Two images of 26+00 used by Großadmiral Dönitz to meet Admiral Arturo Riccardi (centre, second photograph) and other officers of the Regia Marina and Regia Aeronautica at Rome in 1943.

Himmler's personal Fw 200 C-4/U1 after capture at Flensburg in 1945, wearing RAF markings and Air Ministry identity.

Unknown US serviceman photographed in front of abandoned DLH Condor D-AABR Rheinland at Pocking, Bavaria.

Four photographs of W. Nr. 0097 at Bordeaux on a sunny day with KG 40 personnel posing in front of it, with a Peugeot 202 unit car and other personnel evidently enjoying the sunshine. The aircraft wears a code that is not among those usually identified for it, sadly only partially visible with the third and fourth characters either L or E. This aircraft was lost in Nautsi, Finland on 17 October 1944 when it suffered an engine failure during take-off.

CONCLUSION

The story of the Fw 200 Condor is one of effects far out of proportion to the tiny number of airframes, yet still with a sense that far more could have been achieved had the aircraft's full potential been realised.

Vizeadmiral Eberhard Weichold wrote:

At the outbreak of war the Luftwaffe possessed in the trans-ocean aircraft of Lufthansa an excellent means of long distance air reconnaissance. Though these were not war aircraft, and their number was small, yet, if when war started, the type had been further developed and manufactured on a large scale and had been coordinated with the new U-boat building programme, a long-distance reconnaissance service could have been established within a reasonable time. Operating from the French Atlantic bases the aircraft could have surveyed an area up to 1,000 miles in depth in the Atlantic from 40° N to 60° W. It was through this area that the main north–south traffic from the South Atlantic and the east–west North Atlantic traffic to the British isles would have to pass.[86]

The Royal Navy concurred. The Staff History of the Battle of the Atlantic revealed that the Fw 200 threat had been considered equally with that from U-boats in 1941 and noted, "In each battle aircraft assisted the U-boats by their homing procedure. Fortunately, shortage of Condors — there were still only about 20 — prevented consistent exploitation of the technique."[87]

Indeed, it was arguably by acting in concert with U-boats that the greatest damage was done, and the Condor's own anti-shipping activities, while garnering many "Headlines for the Luftwaffe",[88] detracted from the ultimate objective. Nevertheless, for a type of which only 280 were built, the Condor's effects, strategically and psychologically, were astonishing.

---

86   Vice-Admiral E. Weichold, Survey from the Naval Point of View of the Organization of the German Air Force for Operations over the Sea, 1939–45, ONI Review August 1946.
87   E. J. Grove, The Defeat of the Enemy Attack on Shipping 1939–1945, *Naval Staff History* Volume 1A, HMSO 1957, p. 163.
88   The Operational Use of the Luftwaffe in the War at Sea 1939–43, OKL, 8th Abteilung, January 1944, p. 222.

SPECIFICATIONS

FW 200 C

Designation	Werk-Nr	Engine	Fuel load: At maximum range	Fuel load: At highest military payload	Payload: At maximum range (Kg)	Payload: Largest military payload (Kg)	Payload: Largest possible military payload (Kg)	Largest calibre for F	Largest calibre for G	Largest calibre for R	MG in stand A	MG in stand B	MG in stand C	MG in stand D	MG in stand F	Max speed at full power height (Km/H)	Full power height (m)	Maximum range (Km)	Endurance (h)	Service ceiling (m)	Remarks
V-10	0001	132 H-1	7350	—	—	2060	2060			500 (50)						380	1100	4100	13.4	4000	Version "Rowell"
V-11	0002	132 H-1	8060	8060	2060	2060	2060									378	1600	4500	15.8	4000	No repelling loads on the outer wing
C-1	0003/08, 0011/14	132 H-1	8060	4760	1690	4060	4060	1100	500	500 (50)	15	15	15	FF (15)	15	378	1600	4500	15.8	4000	Outward flight 2000 / Return flight 3000
V-12 / C-2	0015/18, 0022/26	132 H-1	8060	2560 (wing tanks)	1690	5400	5400	1800	1400	500 (50)	15	15	15	FF (15)	15	378	1600	4500	15.8	4000	Load aircraft only up to 4500 kg Ln. 3660L. — Outward flight 2000 / Return flight 3000
V-13 / C-3	0025/89, oh 0022..64	323 R-2	8060	2560 (wing tanks)	1290	5400	5400	1800	1400	500 (50)	15	15	15	FF (15)	15	378	1600	4500	14.7	4000	Outward flight 2000 / Return flight 3000
C-3/11	0052	323 R-2	8060	2560 (wing tanks)	1000	5400	5400	1800	1400	500 (50)	15	15	15	151	15	410	4600	4500	14.7	4000	Sample installation of U-Anlage — Delivery condition as C-3
C-3/12	0055	323 R-2	10050	2560 (wing tanks)	500	5150?		1800	1400	500 (50)						410	4600	5250	16.8	4000	Series production of supplemental fuel tanks - Equipment (unprotected, 2 R- lower tub tanks, 2 engine nacelle tanks) unique version, weapon installation, subsequently by KG40
C-3/13	0064	323 R-2	8060	2560 (wing tanks)	—	????		1800	1400	500 (50)	15	131	15	151	15	410	4600	4500	15.7	4000	Series production of the Atlas switchgear, system otherwise as C-3. Single example
C-3/ U4			8060 (standard) / 9310 (with aux tank)	2560 (wing tanks)	1000	4560	4900	1800	1400	500 (50)	151	131	15	151	15						Load aircraft up to 3000 kg — Deflection load with 4760 L. Additionally P one more protected — Deflection load with 5310 L. Additionally 1 more protected R – tub tank. — Load aircraft up to 3000 kg Additionally R – tub tank. LT installation
C-3/15 (C-4)	0070/0094	323 R-2	8060 (standard) / 8600 (with tub tank)	2560 (wing tanks)	500	4560	5400	1800	1400	500 (50)	131	131	15	151	15	410	4600	4560 / 5250	14.7 / 16.8	4000	Load up to 2500 kg Deflection load with 5310 L. LT installation — Load aircraft up to 4800 kg R – tub tank. LT installation — Deflection load with 4060 kg
C-3/16	From 0070 to 0094	323 R-2	7060	2560 (wing tanks)	—	5360	5460	1800	1400		131	131	15	151	15	410	4600	4560	14.7	4000	Conversion of A-Stand from C-3/U-4 variant undertaken by KG40 (MG 131)
C-3/17	oh 0095	323 R-2	9140	2560	—	4400	4900				151	131	15	FF (15)	15	410	4600			4000	Special version (no further details given)
C-3/18	Before 0070	323 R-2	9140	2560	1000	5400	5400				131	131	15	151	15	410	4600	5400	16.8	4000	Conversion of A-stand from C-3/U-4 undertaken by KG40 (MG 15 replaced with MG 131)
C-3/19	0099	323 R-2	3660	2560	—						131	131	15	151	15	410		2030 (10% reserve)	6.7	4000	government aircraft: 11 passengers 1 fuselage auxiliary tank.

Interior view of the improved D-stand with 'lens' gun mount, showing the Lofte 7D bombsight – sliding shutters covered the sight when it was not in use.

The cockpit of the V1 and a comparison shot of a C-3 cockpit showing how little the layout changed throughout the production run. The main difference is the Revi gunsight on the military version, generally used for low-level bomb attacks.

Two views of the direction-finding aerial in the nose of all Condors until the radar-equipped variants.

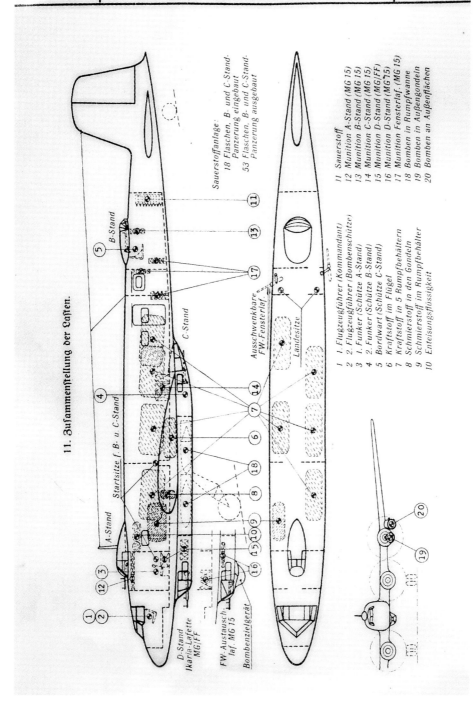

Diagram of the Fw 200 C-1 showing the layout of the fuel tanks, defensive guns and bomb carriage.

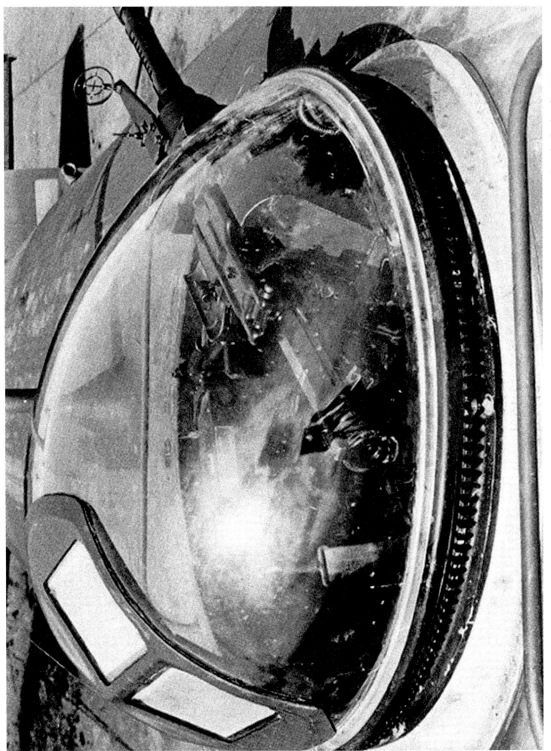

The low-profile D30 gun mount introduced in the A-stand of the C-3 model and used in numerous variants since, even after the hydraulic HD151 was introduced. It was up-gunned in 1942 from the MG 15 to the MG 131.

KENNBLATT FUER FLUGZEUGMUSTER FW 200 C-1 UND C-2 SERIE MIT BMW 132 H 1 MOTOREN	DATASHEET FOR AIRCRAFT TYPE FW200 C1 AND C2 SERIES WITH BMW 132 H1 MOTORS
Verwendungszweck: Hilfsbomber mit 5 MG 15, 2 MG/FF u. maximal 4000kg Bombenlast. Fernerkunder	Purpose: Auxiliary bomber with 5 M15, 2 MG/FF, max 4,000kg bomb load, long-distance reconnaissance
Musterbezeichnung: Fw 200 C-1, C-2	Type designation FW 200, C-1, C-2
Besatzung: 5 Mann	Crew: 5 men
Bauweise: 4 motoriger freitragender Tiefdecker in Ganzmetallbauweise mit einziehbarem Fahrwerk. Glattblechbeplankung **Raumaufteilung:** Bugklappe aus Holz mit Peilrahmen **Fuehrerraum** 2 Fuehrersitze nebeneinander, hinter dem rechten Fuehrersitz Navigationstisch, hinter dem linken Fuehrersitz bequemer Ruhesitz, quer zur Flugrichtung **Funkerraum:** links vorne FuG X – Geraete, links hinten DLH Lorenz Kurzwellen Station, rechts unten Munition fuer den vorderen Stand, rechts oben Ablageboerder fuer Lebensmittel und Kochplatte Oben: vorderer Schuetzenstand (A-Stand) Unten: Durchstieg zum vorderen unteren Stand (D-Stand) **Kleiner Rumpfraum:** Rechts vorn Oelnachtankbehaelter, rechts hinten Schalttisch fuer Kraft und Schmierstoff anlage Links vorn Navigationsstisch, links oben 3 Behaelter fuer Hydraulikoel, Luftschrauben und Vergaser Enteisungsfluessigkeit 1 Kraftstoffbehaelter **Grosser Rumpfraum:** Rechte Seite vorn nach hinten angeordnet: Schlauchboot, 1 Kraftstoffbehaelter, Durchstieg zum hinteren unteren Stand (C-Stand), 3 Trommel Magazin Behaelter, 3 Maschinenpistolen mit Munition, 1 Kraftstoffbehaelter, Navigationstisch, darunter Bildgeraet Linke Seite: 2 Kraftstoffbehaelter, 1 Reserve MG 15, Einstiegstuer **Hinterer Zwischenraum:** Rechts und Links 1 ausschwenkbare Fensterlafette mit MG15 Am Boden 2 Liegematratzen **Hinterer Rumpfraum:** Hinterer oberer Stand (B-Stand), Bildgeraet, Sauerstoffflaschenbatterie, Toiletteneimer	Construction: 4 engine cantilever low-wing plane in full metal construction with retractable landing gear. Smooth sheet metal skin **Room layout:** Wooden nose cone with sighting frame **Cockpit:** 2 pilot seats next to each other, behind the right pilot seat navigation table, behind left pilot seat comfortable resting seat transverse to direction of flight **Radio Operator's cabin:** front left FuG X devices, back left DLH Lorenz short wave station, bottom right ammunition for front shooting stand, top right shelving for food and cooking hotplate Top: front gun position (A-Stand) Bottom: access to front lower gun position (D-Stand) **Small fuselage space:** Front right: oil-refuelling container Back right: switchboard for fuel and lubricant system Front left navigation table, top left 3 containers for hydraulic oil, rotor and carburettor de-icing fluid, 1 fuel container **Main fuselage space:** Right side front to back positioned: inflatable boat, 1 fuel tank, access to lower rear gun position (C Stand), 3 drum magazine containers, 3 machine guns with ammunition 1 fuel container, navigation table, below that imaging device Left side: 2 fuel tanks, one reserve MG 15, access door **Rear intermediate fuselage space:** Right and left 1 swing-out gun mount with MG 15 On the floor two mattresses **Rear fuselage space:** Rear upper gun position (B Stand), imaging device, oxygen tank battery, Toilet bucket
Rumpf: Schalenrumpf mit senkrecht zur Laengsachse angeordneten Spanten. Versenkt genietete Glattblechaussenhaut	**Fuselage:** monocoque with frames that are perpendicular to the longitudinal axis Smooth metal skin
Rumpfwanne: Unter dem Rumpf Wanne in Schalenbauart fuer Bombenraum und C und D Stand	**Fuselage tub:** underslung fuselage tub in shell construction type for bomb space and C and D stand

Tragwerk: 3teiliger 2 holmiger Fluegel. Mittelteil das den Rumpf, die 4 Triebwerke und das Fahrgestell traegt und je 1 Aussenfluegel hinter dem Hauptholm mit Stoff bespannt.	**Wing unit:** 3 part 2 spar wing centre unit carrying the fuselage, 4 engines and the landing gear and 1 outer wing each covered with fabric behind main spar
Gondeln: Am Fluegelmittelteil je 2 unter jeder Seite. Vorderteil mit Vorbau vor der Flaeche fuer die Triebwerke, hinterer Teil bei den inneren Gondeln fuer Einziehfahrwerk, bei den aeusseren als Bombenraum ausgebildet **C 1** – Alle Gondeln durch Klappen abgeschlossen **C2** – Aeussere Gondeln offen	**Nacelles:** at wing centre unit 2 each under each side. Front part with cowling in front of the wing for engines, back part of inner nacelles for retractable landing gear, in outer as bomb space **C1**: all nacelles shut with flaps **C2**: outer nacelles open
Leitwerk: **Querruder** – 2 teilig, etwa ueber ¾ der Spannweite des Aussenfluegels mit Gewichtsausgleich. Der hintere Teil mit Stoff bespannt. Sich selbst einstellende Hilfsruder. Das linke inner Hilfsruder als Trimmruder mit Elektromotor vom Fuehrerraum aus zu betaetigen **Hoehenleitwerk:** Hoehenflosse, 2 holmig mit Duralblech beplankt. Verstellmoeglichkeit am Boden vorhanden Hoehenruder, 2teilig mit einem Holm. Bis Holm Duralblech beplankt, hinter Holm Stoff bespannt. Vollstaendig gewichtsausgeglichen. Selbsttaetig gesteuerte Hilfsruder. Linkes Ruder mit elektrisch angetriebenem Trimmruder mit Betaetigung vom Fuehrerhaus aus. **Seitenleitwerk:** Seitenflosse, 2 holmig bis hinterholm mit Duralblech beplankt, ab Hinterholm stoffbespannt. Seitenruder einholmig mit Duralblech beplankt. Vollstaendig gewichtsausgeglichen, Luftkraftausgleich durch ein selbstaetig gesteuertes Hilfsruder und durch ein elektrisch angetriebenes Trimmruder, vom Fuehrerraum zu betaetigen **Landeklappen:** Spreizklappen ueber die ganze Laenge des Fluegelmittelstueckes ausser der Rumpfbreite und etwa ¼ des Aussenfluegels. Aufbau 2 holmig mit Elektron Blech beplankung	**Controls** **Aileron:** 2 part, along ¾ of outer wingspan with mass balance. Back part covered with fabric. Self-adjusting (automatic?) auxiliary rudder as trim rudder with electric motor operated from cockpit **Tailplaine:** tailplane, 2 spars with duralumin skin. Option to adjust on the ground available Elevator: 2 part with 1 spar, duralumin skin up to spar. Behind spar covered with fabric Completely counterbalanced. Automatic auxiliary rudders, left rudder with trim rudder with electric motor that is operated from cockpit **Vertical tail:** fin 2 spars, duralumin skin up to rear spar, after back spar fabric covered Rudder single spar with duralumin skin, completely counterbalanced Aerodynamic force compensation via automatic auxiliary rudder and a electric powered trim rudder that is operated from the cockpit **Landing flaps:** cowl flaps across the full length of the wing centrepiece with exception of the width of the fuselage and ca ¼ of the outer wing. Construction 2 spars with Elektron skin.[89]
Steuerung: Doppelsteuer vom 1. Und 2. Fuehrer gleichzeitig zu bedienen. Rechte Steuersaeule auskuppelbar Hoehen und Seitensteuerrung mit Differential Steuerrung zwecks Herabsetzung der Steuerkraefte	**Flying controls:** dual controls to be operated by 1st and 2nd pilot simultaneously. Right hand controls can be disengaged Altitude and lateral control with differential action to lower control forces
Fahrwerk: Einziehfahrwerk hydraulisch vom Fuehrer ein und ausfahrbar. Je Fahrwerkshaelfte mit 2 Laufraedern, bremsbar,Notbetaetigung durch electromotor betriebenes Notaggregat Sporn: Einziehbares Spornrad, allseitig drehbar, mit Daempfung gegen Flattern beim Rollen	**Landing gear:** retractable landing gear, retractable by pilot via hydraulics. Each half has 2 wheels, can be braked, emergency operation via electric emergency generator. Tail: retractable tail wheel, can swivel in all directions, shock absorption against shimmy when rolling

89 Elektron is an alloy made of 90% magnesium and 10% aluminium, and traces of zinc and tin which is resistant to corrosion

Hydraulikanlage: hydraulisch betaetigt werden Fahrwerk und Sporn. Landeklappen, Fahrwerksbremsen, Betrieben durch je 1 Oelpumpe and den beiden Innenmotoren 1 Oelbehaelter mit einfachem Beschussschutz Inhalt 35l, Einfuellmenge 27l Druckoel: Shell AB 11	**Hydraulics:** Hydraulically operated landing gear and spur, landing flaps, landing gear brakes. Operated via 1 oil pump each and on both inner motors 1 oil container with basic armour, 35l, filling level 27l Hydraulic oil: Shell AB11
Enteisung: **Gummienteiser** am Leitwerk **Luftschrauben-Enteisung** durch Enteisungsfluessigkeit 1 Behaelter mit einfachem Beschussschutz, Inhalt 35l, Einfuellmenge voll **Enteisungsfluessigkeit:** LEF 25 A **Vergaser-Enteisung** durch Warmluft und Spirituszusatz zur Vergaseransaugluft 1 Behaelter mit einfachem Beschussschutz Inhalt 35l, Einfuellmenge voll **Enteisungsfluessigkeit:** VEF 5 **C-2 Warmluftenteisung** an den Aussenflaechennasen	**De-icing:** **Rubber de-icer** at flying controls **Propeller de-icing** using de-icer fluid 1 container with basic armour, 35l, filling level full De-icer fluid: LEF 25 A **Carburettor de-icing** via warm air and alcohol additive to carburettor inlet air 1 container with basic armour 35l, filling level full **De-icer fluid:** VEF 5 **C-2 Warm air de-icing** at outer plane noses
Panzerung Fuehrersitz links 8 mm Panzerplatte vorn, unten und hinten mit Kopfschutz **A-Stand** nicht geschuetzt **B-Stand** 8mm Panzerplatte hinten Panzerkragen oben hinten **C-Stand** 8mm Panzerplatte unten **D-Stand** 8mm Panzerplatte unten nur bei MG/FF Lafette Panzerung insbesondere B-und C Stand Panzerung leicht ausbaubar	**Armour:** pilot seat left 8mm armour plate front, bottom and back, with head protection **A Stand** no armour **B Stand** 8mm armour plate back and armoured collar top back **C Stand** 8mm armour plate bottom **D Stand** 8mm armour plate bottom, only with MG/FF gun carriage Armour, especially B and C stand armour, easily removable

Triebwerk
Motorenmuster BMW 132 H1
Verdichtung E = 6,5

1 Min Leistung
Nmax 4 x 1000 PS
n max = 2550 U/min
p = 1,42 ata
Kuehler –
Luftschrauben
3 flgl. VDM Verstell Metall Luftschraube
D = 3,5m
Grundeinstellung: 25"
Schraubenuhrzeit: 1116
Startstellung: 12°°.
Kraftstoffbehaelter: voll geschuetzt
4 x Behaelter im Fluegel je 380 l = 1520 l
4 x Behaelter im Fluegel (Startbehaelter) je 260 l = 1040l
5 x Behaelter im Rumpf je 1100l = 5500l
Gesamt = 8060 l
Nicht ausfliegbare Restmenge = 260l
Betankung
Fluegelbehaelter von Fluegeloberseite
Rumpfbehaelter durch Aussenbordanschluesse 3 links 2 rechts an der Rumpfaussenseite oberhalb der Fluegel
Kraftstoff Schalt und Pumpanlage
Rumpfbehaelter auf alle Triebwerke schaltbar
Kraftstoffoerderung bis Werk Nr 0007 durch 2 motorangetriebene Jumo Kraftstoffpumpen, ab Werk Nr 0008 durch Graetzin Kraftstoffpumpen.
Kraftstoffentnahme in grossen Flughoehen mit Behaelterpumpen in jedem Kraftstoffbehaelter
Schnellablass –
Schmierstoffbehaelter Einfacher Beschussschutz
4 x Behaelter in den Triebwerksvorbauten je 36 l = 144l
1 x Behaelter im Rumpf (Nachtankbehaelter) = 450l
Gesamt 594l

Fuellinhalt: 4 x 30 l = 120l
1 x 450 l = 450 l
Gesamt = 570 L
Betankung
Betriebsbehaelter von oben
Nachtankbehaelter durch Aussenbordanschluss rechts an der Rumpfaussenseite oberhalb des Fluegels
Nachtankanlage
Handgetriebene Allweiler Pumpe am Schalttisch im kleinen Rumpfraum
Schmierstoffregler
Bauart "Bruhn" Frischoelgefeuerter Ventil Durchfluss regler mit 2 Membrankoerper im Druckausgleich
Zu verwendender Schmierstoff
Intava 100 oder ASM Rotring
Kaltstartanlage vorhanden
Flammendaempfer vorhanden

Powerplant
Engines BMW 132 H1
Compression E = 6,5

1 Min Performance
Nmax 4 x 1,000 PS
n max = 2550 U/min
p = 1.42 ata
Propellers
3 blade. VDM adjustable metal rotors
diameter = 3.5m
basic configuration: 25"
Rotor time: 1116
Start position: 12°°.
Fuel tank: fully armoured
4 x containers in wing, each 380l = 1520l
4 x containers in wing (starting containers) each 260 l = 1040l
5 x containers in fuselage each 1,100l = 5,500l
Total = 8,060l
Amount below which tanks cannot fall = 260l

Fuelling
Wing tanks fuelled from top of wing
Fuselage tanks fuelled via outboard connection 3 left 2 right on outside of fuselage above wing
Fuel switch and pump system
Fuselage container can be switched to all engines. Fuel pump up to works no 0007 via 2 motorised Jumo fuel pumps, from works no 0008 via Graetzin fuel pumps.
Fuel extraction in high altitudes with container pumps in each fuel container
Quick jettison
Lubricant container basic armour
4 x containers in engine nacelles each 36l = 144l
1 x container in fuselage (for refuelling) = 450l
Total 594l

Filling level: 4 x 30l = 120l
1 x 450l = 450l
Total = 570L
Fuelling
Active container from top
Refuelling container r via outboard connection on right of fuselage outside above wing
Refuelling system
Hand operated Allweiler Pump at switch board in small fuselage room
Lubricant control
Construction type 'Bruhn fresh oil-fired valve flow regulator with 2 membrane bodies in pressure equalisation
Lubricant to be used
Intava 100 oder ASM Rotring
Cold start system existing
Exhaust flame damper existing

Ausruestung und Geraete Die Ausruestung und Geraete sind aus den Ag – Listen zu entnehmen. Sie bestehen aus:

a) Flugueberwachungsgeraete
Fuer Nacht und Blindflug ausreichend

b) Sicherheits – und Rettungsgeraete
Sauerstoffanlage
53 Flaschen bei ausgebauter B und C Stand Panzerung
18 Flaschen bei eingebauter B und C Stand Panzerung
8 Atemgeraete
2 tragbare Geraete
5 Schnellklinkfallschirme,
5 Fallschirmhalterungen dazu
Rechter und Linker Fuehrersitz und Funkersitz fuer Sitzkissenfallschirm
8 Sitze mit Bauchgurt
A Stand mit Sitzgurt
5 Halterungen fuer Gasmasken
1 Sanitaetspack
1 Sanitaetstasche
1 Schlauchboot fuer 6 Personen
Feuerloeschanlage
2 Feuerloescher fuer Triebwerke
2 Handfeuerloescher im Rumpf

c) Nachrichten und Verstaendigungsgeraete
Bordfunkgeraet: FuG X
Peilgeraet: Peil G 5
Blindlandeanlage: FuBl 1,
Eigenverstaendigung: Ei V
Zusatzfunkgeraet: DLH Lorenz- Kurzwellenstation
Leuchtpistole mit 12 Schuss
d) Navigationsgeraete
Patin-Kompassanlage
Kompl. Aursuestung fuer astron. Navigation
e) Kurssteuerung
Askania-Letz 14a
f) Bewaffnung
A Stand Llg Lafette mit MG 15, 1125 Schuss, wird nachtraeglich geaendert in Drehkranz D30 (360° schwenkbar) mit MG15 und Plexiglashaube
B-Stand Drehkranz 30 mit Rittervisier und MG 15, 1125 Schuss
C- Stand Kegellafette mit MG15, 1125 Schuss
C1: D-Stand Ikaria Lafette mit MG/FF, 300 Schuss, MG/FF austauschbar gegen MG 15 mit beschraenktem Schussbereich bei Bombenzielgeraeteeinbau, 1125 Schuss
C2: D-Stand Ikaria Lafette mit MG/FF, 300 Schuss, austauschbar gegen Focke-Wulf Fensterlafette rechts und links mit MG 15, 1125 Schuss
Je 1 ausschwenkbare Fode Wulf Fensterlafette rechts und links mit MG 15, 1500 Schuss
1 Reserve MG 15,
3 Maschinenpistolen 450 Schuss

Equipment and devices equipment and devices to be taken from Ag/ lists. They consist of:

a) Flight instruments sufficient for night and blind flight
b) Safety and rescue devices
Oxygen system
53 bottles if B and C stand armour removed
18 bottles if B and C stand armour present
8 breathing devices
2 portable devices
5 fast jack parachutes
5 parachute mountings and additionally right and left pilot seat and radio operator seat seating parachute
8 seats with waist belt
5 gas mask mountings
1 first aid kit
1 first aid bag
1 inflatable boat for 6 persons
2 fire extinguishers for engines
2 manual fire extinguishers in fuselage

c) Communication devices
On-board radio: FuG X
Direction finder: Peil G5
Blind landing system: FuBl 1
Intercom: Ei V
Additional radio: DLH Lorenz short wave station
Flare gun with 12 shots
d) Navigation equipment
Patin compass system
Complete equipment for astronomical navigation
e) Course control
Askania Letz 14 a
f) Weaponry
A Stand Llg mount with MG 15, 1,125 rounds, subsequent adjustment in slewing ring D30 (360° pivot) with MG 15 and plexiglass cover
B Stand slewing ring 30 with 'knight's visor' and MG 15, 1,125 rounds
C Stand cone mount with MG 15 1,125 rounds
C-1 D Stand Ikaria mount with MG/FF 300 rounds, exchangeable for MG 15 with restricted shooting field when bombing aim device included
C-2 D Stand Ikaria mount with MG/FF 300 rounds, exchangeable for Focke-Wulf window mount left and right with MG 15, 1,125 rounds
Each 1 swing out window mount right and left with MG 15, 1,500 rounds
1 reserve MG 15
3 machine guns 450 rounds

<table>
<tr><td>

g) Bombenauruestung
Bombentraeger:
Rumpfwanne: 3x4 Schloss 50 oder
2 x ETC 500 IX b.
C1: Auessere Motorengondeln 2 x ETC 500 IXb, 2 PVC
1006 L.
C2: Aussenfluegel: 2 x PVX 1006 L

Bombenbelademoeglichkeit:
Rumpfwanne 12 x SC 50 oder 12 x SD 50 oder
12 Leuchtfallschirmbomben oder
2 x SC250 oder 2 x SC500 oder 2 x SD500 oder 2 x PC500
C1: Aeussere Motorgondeln 2 x SC250 oder 2 x SC500
oder 2 x SD500 oder 2 x PC500 oder 2LMA
C2: Wie C1, ausserdem 2 x SC1000 oder 2 x SD1000 oder 2
x PC 1000 oder 2 x SD1400 oder 2 x LMB
C1 Aussenfluegel: 2 x SC 250 oder 2x SC500 oder 2 x
SD500 oder 2 x PC500 oder 2 x SC1000 oder 2 x PC1000
oder 2 x LMA oder 2 x LMB
C2: wie C1 ausserdem 2 x SD1400 oder 2 x SC1700 oder 2
x PC1700 oder 2 x SC 1800

Groesste Bombenlast bei vollen Kraftstoffbehaeltern:
1630kg
Groesstmoegliche Bombenlast
C1: 4060kg
C2: 4900kg

Ausloeseanlage:
RAB 14c in Rumpfwanne (DStand)
B-Knopf XI am linken Steuerhorn
Zuenderanlage:
Zuenderanlage fuer Sprengmunition ZZG 1/24-Einbau fuer
Zeitzuender

a) Zielgeraete
C1: GV219d in Ikaria Lafette oder Lotse 7c in Ikaria Lafette
behelfsmaessig
Revi C/12C im Fuehrerraum Tiefanflug
C2: GV 219d in Ikaria Lafette oder Lotse 7b in FW-
Austauschlafette oder Lotse 7c in FW Austauschlafette oder
BZG 2 L in FW –Austauschlafette
Revi C/12C in Fuehrerraum fuer Tiefanflug
b) Bildgeraete
Reihenbildner Rb 50/30
Reihenbildner RB 20/30
Handkamera 12,5/7 X 9

</td><td>

g) Bombing equipment
bomb carrier
fuselage tub 3x4 Schloss 50 or
2 x ETC 500 IX b.
C1: outer nacelle 2 x ETC 500 IXb, 2 PVC 1006 L.
C2: outer wing: 2 x PVX 1006 L

Bomb loading options:
Fuselage trough 12 x SC 50 or 12 x SD 50 or
12 parachute flares or
2 x SC250 or 2 x SC500 or 2 x SD500 or 2 x PC500
C1: outer nacelles 2 x SC250 or 2 x SC500 or 2 x SD500 or
2 x PC500 or 2LMA
C2: as C1, additionally 2 x SC1000 or 2 x SD1000 or 2 x PC
1000 or 2 x SD1400 or 2 x LMB

C1 outer wing: 2 x SC 250 or 2x SC500 or 2 x SD500 or 2
x PC500 or 2 x SC1000 or 2 x PC1000 or 2 x LMA or 2 x
LMB
C2: as C1 additionally 2 x SD1400 or 2 x SC1700 or 2 x
PC1700 or 2 x SC 1800

Max bombload fully fuelled: 1,630kg
Max overall bomb load
C1: 4,060kg
C2: 4,900kg

Trigger
RAB 14c in fuselage (DStand)
B-Knopf XI on left steering horn
detonator
detonator for explosives ZZG 1/24-installation for time
fuse

a) Target device
C1: GV219d in Ikaria mount or Lofte 7c in auxiliary Ikaria
mount
Revi C/12 C in cockpit for Low altitude
C2: GV 219d in Ikaria mount or Lofte 7b in FW-
replacement mount or Lofte 7c in FW replacement mount
or
BZG 2 L in FW-replacement mount
B b) Cameras
Reihenbildner Rb 50/30
Reihenbildner RB 20/30
Handkamera 12,5/7 X

</td></tr>
<tr><td>

Sondereinrichtung
Nachtflugtauglich.
2 x 3 kg Flugzeug-Zerstoerkoerper
1 Abtristmesser Heyde M17
2 Luft Liegematratzen
8 Anschluesse fuer Heizbekleidung

</td><td>

Extra equipment
Night flight able
2 x 3kg airplane destroyer
1 x ?
2 Airbeds
8 connections for heatable clothes

</td></tr>
</table>

Transportfaehigkeit	**Transportation**
Eisenbahntransport 8 Wagen	Transport by rail, 8 carriages
1 SSL Wagen 18m fuer Rumpf	1 SSL carriage 18m for fuselage
1 SSL Wagen 18m fuer Innenfluegel	1 SSL carriage 18m for inner wing
1 SSL Wagen 14m fuer Rumpfhinterteil, 4 Triebwerke	1 SSL carriage 14m for back part of fuselage, 4 engines
2 R-Wagen 11 m fuer Aussenfluegel	2 R-carriages 11m for outer wing
1 Glt Wagen 10,8m fuer Leitwerke und Rumpfwanne	1 Glt carriage 10.8m for rudders and fuselage trough
1 G Wagen 8,3m fuer Fahrwerke, Querruder und (sic) Ladeklappen	1 G carriage 8.3m for landing gear, aileron and landing flaps
1 G Wagen 8,3m fuer Betriebstoffbehaelter und Luftschrauben	1 G carriage 8.3m for fuel tanks und propellers
Abmessungen	**Dimensions**
Laenge des Flugzeuges 23,85m	Length of airplane 23.85m
Hoehe des Flugzeuges mit Antennenmast 6,30m	Height of plane including antenna 6.30m
Spannweite der Tragflaechen 32,84m	Wingspan 32.84m
Tragflaeche insgesamt 118 m2	Wing area in total 118m2
Technische Daten	**Technical data**
Beanspruchungsgruppe H3	Use group H3
Hoechszulaessiges Abfluggewicht 22 700kg	Max. permissible take-off weight 22,700kg
Landegewicht normal 17 500 kg	Normal landing weight 17,500kg (without bombs, 2/3 fuel
(ohne Bomben, 2/3 Kraftstoff leer)	empty), max permissible landing weight 19,000kg
Hoechstzulaessiges Landegwicht 19 000kg	Max permissible speed:
Hoechstzulaessige Geschwindigkeiten	Gliding
Gleitflug	0–3,000m altitude va 440km/h
In 0-3000m Hoehe va 440km/h	Above 3,000m altitude va 400km/h
Ueber 3000m Hoehe va 400km/h	Bad weather near ground va 230km/h
Bei unsichtigemWetter in Bodennaehe va 230km/h	3 engine flight
3 -Motorenflug	If G = 21,0 t (bombs discharged)
Bei G= 21,0t (Bombenlast abgeworfen)	Service ceiling 2,000m
Dienstgipfelhoehe 2000m	Cruising speed va = 240km/h
Reisegeschwindigkeit va= 240km	Max speed in 1,000m altitude
Hoechstgeschwindigkeit	Without external loads va= 350km/h
In 1000 m Hoehe	2 motor flight
Ohne Aussenlasten va= 350km	G= 16,5t 30 min Motor performance,
2 Motoren Flug	Able to fly (according to FW) at 1,400m altitude
G= 16,5t 30 min Motorleistung,	Wing load at starting weight = 22.7t 192kg/m3
Flugfaehig (nach Angabe FW) in 1400 m Hoehe	Not suitable for catapult
Flaechenbelastung bei	
Abfluggewicht = 22,7 t 192kg/m3	
Nicht katapultfaehig	

INDEX